Study Guide

for

Pytel and Kiusalaas's

Engineering Mechanics: Dynamics

Third Edition

Andrew Pytel

Pennsylvania State University

Jean Landa Pytel

Pennsylvania State University

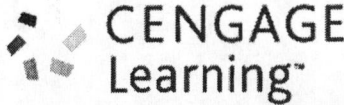

Australia • Brazil • Japan • Korea • Mexico • Singapore • Spain • United Kingdom • United States

Study Guide for Engineering Mechanics: Dynamics, 3e

Andrew Pytel
Jean Landa Pytel

Director, Global Engineering Program:
Chris Carson

Senior Developmental Editor:
Hilda Gowans

Editorial Assistant:
Jennifer Dinsmore

Marketing Specialist:
Lauren Betsos

Director, Content and Media Production:
Renate McCloy

Production Service:
RPK Editorial Services

Senior Art Director:
Michelle Kunkler

Cover Designer:
Andrew Adams

Cover Image:
iStockphoto/Olga Paslawska

Senior Manufacturing Coordinator:
Joanne McNeil

© 2010, 1999 Cengage Learning

ALL RIGHTS RESERVED. No part of this work covered by the copyright herein may be reproduced, transmitted, stored, or used in any form or by any means—graphic, electronic, or mechanical, including but not limited to photocopying, recording, scanning, digitizing, taping, Web distribution, information networks, information storage and retrieval systems, or in any other manner—except as may be permitted by the license terms herein.

> For product information and technology assistance, contact us at
> **Cengage Learning Customer & Sales Support, 1-800-354-9706.**
>
> For permission to use material from this text or product, submit all requests online at **www.cengage.com/permissions**. Further permissions questions can be emailed to
> **permissionrequest@cengage.com**

Library of Congress Control Number: 2008943459

U.S. Student Edition:

ISBN-13: 978-0-495-41124-6

ISBN-10: 0-495-41124-8

Cengage Learning
200 First Stamford Place, Suite 400
Stamford, CT 06902
USA

Cengage Learning is a leading provider of customized learning solutions with office locations around the globe, including Singapore, the United Kingdom, Australia, Mexico, Brazil, and Japan. Locate your local office at: **international.cengage.com/region**.

Cengage Learning products are represented in Canada by Nelson Education Ltd.

For your course and learning solutions, visit
www.cengage.com/engineering.
Purchase any of our products at your local college store or at our preferred online store **www.ichapters.com**.

Printed in [the United States of America]
1 2 3 4 5 6 7 12 11 10 09

PREFACE TO THE STUDENT

This *Study Guide* is written to accompany *Dynamics*, 3e, Pytel and Kiusalaas, 2010. The sole purpose of this *Study Guide* is to help you master the fundamentals of engineering dynamics as presented in Chapters 11–18 in the textbook. This *Study Guide* is intended to supplement the textbook, not replace it.

There are twenty-nine lessons, with each lesson devoted to a particular reading assignment in your textbook. Numerical methods and optional topics are not included.

Each lesson contains the following two parts:

A. SELF-TEST

The Self-Tests review and reinforce the fundamental principles presented in the reading assignment in your textbook. You should take each Self-Test immediately after you have completed the assigned reading, and before solving the Guided Problems contained in the lesson. The answers to the Self-Tests are given in Appendix A.

B. GUIDED PROBLEMS

The Guided Problems give you the opportunity to work through the solution of one or more problems before you attempt to solve the homework problems. As the name suggests, the unique feature is that you are "guided" through the solutions of a representative problems. Working through the "fill-in-the-blanks" format for the solutions will help prepare you to solve the homework problems. The solution to each Guided Problem is presented in Appendix B.

A WORD ON NOTATION

The notations used in the lessons follow the conventions used in the textbook: (1) scalars are written as italicized English or Greek letters (such as t for time and θ for an angle); (2) vectors are written as bold-faced letters (such as **F** for force); and (3) the magnitude of a vector **A** is denoted by |**A**| or simply as A. In our hand-written solutions in Appendix B, we use an arrow above a symbol to indicate that the symbol represents a vector quantity. For example, $\vec{A}$ (handwritten) refers to the vector **A**. Of course, you should use the notation for vectors with which you are comfortable. However, it is important that you clearly, and consistently, indicate the symbols that represent vector quantities.

CONTENTS

PREFACE   page

    Lesson 1   Introduction to Dynamics.. 1

Dynamics of Particles

    Lesson 2   Kinematics: Rectangular Coordinates... 3
    Lesson 3   Dynamics of Rectilinear Motion: Force-Mass-Acceleration Method................. 9
    Lesson 4   Curvilinear Motion.. 17
    Lesson 5   Kinematics: Path (Normal-Tangential) Coordinates..................................... 21
    Lesson 6   Kinematics: Polar and Cylindrical Coordinates.. 26
    Lesson 7   Force-Mass-Acceleration Method: Curvilinear Coordinates........................... 29
    Lesson 8   Work of a Force; Principle of Work and Kinetic Energy............................... 33
    Lesson 9   Conservative Forces and Conservation of Mechanical Energy........................ 39
    Lesson 10   Power and Efficiency.. 42
    Lesson 11   Principle of Impulse and Momentum.. 44
    Lesson 12   Principle of Angular Impulse and Momentum... 47

Dynamics of Systems of Particles

    Lesson 13   Kinematics of Relative Motion... 50
    Lesson 14   Kinematics of Constrained Motion.. 52
    Lesson 15   Kinetics: Force-Mass-Acceleration Method.. 54
    Lesson 16   Work-Energy and Impulse-Momentum Principles....................................... 57
    Lesson 17   Plastic Impact and Impulsive Motion... 65
    Lesson 18   Elastic Impact... 70

Plane Motion of Rigid Bodies

    Lesson 19   Plane Angular Motion; Rotation about a Fixed Axis................................... 73
    Lesson 20   Relative Motion; Method of Relative Velocity... 77
    Lesson 21   Instant Centers for Velocities... 84
    Lesson 22   Method of Relative Acceleration.. 88
    Lesson 23   Absolute and Relative Derivatives of Vectors; Rotating Reference Frames.......... 92
    Lesson 24   Mass Moment of Inertia; Composite Bodies.. 101
    Lesson 25   Angular Momentum of a Rigid Body; Equations of Motion; FMA Method.......... 104
    Lesson 26   Work and Power of a Couple; Kinetic Energy of a Rigid Body........................ 113
    Lesson 27   Work-Energy Principle and Conservation of Mechanical Energy..................... 118
    Lesson 28   Momentum Diagrams; Impulse-Momentum Principles................................. 121
    Lesson 29   Rigid-Body Impact.. 126

APPENDIX A: ANSWERS TO SELF-TESTS.. 131

APPENDIX B: SOLUTIONS TO GUIDED PROBLEMS.. 139

Lesson 1 Introduction to Dynamics
Text Reference: Chapter 11

Note: This lesson contains only a Self-Test; there are no Guided Problems.

A. SELF-TEST (*To be done after assigned reading has been completed.*)

1. A body with *negligible dimensions* is referred to as a _____.

2. A body that undergoes *negligible deformation* is referred to as a _____.

3. The branch of dynamics that studies the geometry of motion, without considering the forces that cause the motion, is called _____.

4. The branch of dynamics that relates the forces to the motion is called _____.

5. Motion that is described with respect to a fixed coordinate system is known as (a) _____ motion. Motion that is described with respect to a moving coordinate system is known as (b) _____ motion.

6. List the three main methods of kinetic analysis:
   _____ ; _____ ; _____

7. A vector **A** is a function of a scalar parameter t: **A**(t). Write the expressions for the following:

   (a) the magnitude of the derivative of **A** with respect to t:

   (b) the derivative of the magnitude of **A** with respect to t:

8. The vector drawn from a fixed origin to a moving point is called the _____ of the point.

(continued)

9. Using **r**(*t*) as the position vector of a particle (*t* is time), write the definition for:

    (a) the velocity **v**(*t*) of the particle in terms of its position vector **r**(*t*):

    v(*t*) =

    (b) the acceleration **a**(*t*) of the particle in terms of its velocity **v**(*t*):

    a(*t*) =

    (c) the acceleration **a**(*t*) of the particle in terms of its position vector **r**:

    a(*t*) =

10. During a certain time interval Δt, a particle moves from point *A* to point *B*. The vector drawn from *A* to *B* is called the _____ of the particle during the time interval.

11. (a) The velocity vector is *always* tangent to the path of a particle. (T or F) ____
    (b) The acceleration vector is *always* tangent to the path of a particle. (T or F) ____

12. How is the speed of a particle related to its velocity?
    _____

13. What is the dimension of the unit "pounds per square inch"? _____

Lesson 2 Kinematics: Rectangular Coordinates

Text Reference: Articles 12.1 and 12.2; Sample Problems 12.1–12.4

A. SELF-TEST (*To be done after assigned reading has been completed.*)

1. A particle moves in a fixed rectangular reference frame. Fill in the missing terms in each of the following vector expressions, where x, y, and z are time-dependent rectangular coordinates of the particle:

   (a) Position vector of the particle:

   $$\mathbf{r} = x\mathbf{i} + \underline{\qquad} + \underline{\qquad}$$

   (b) Velocity vector of the particle:

   $$\mathbf{v} = v_x\mathbf{i} + \underline{\qquad} + \underline{\qquad} = \dot{x}\mathbf{i} + \underline{\qquad} + \underline{\qquad}$$

   (c) Acceleration vector of the particle:

   $$\mathbf{a} = a_x\mathbf{i} + \underline{\qquad} + \underline{\qquad} = \ddot{x}\mathbf{i} + \underline{\qquad} + \underline{\qquad}$$

2. What is meant by *coplanar* motion of a particle?

   _____

3. What is meant by *rectilinear* motion of a particle?

   _____

4. A particle is undergoing motion in the *yz*-plane. If you were going to use the equations listed in Question 1 above, what modifications would you make?

   _____

5. A particle is undergoing motion along the *z*-axis. If you were going to use the equations listed in Question 1 above, what modifications would you make?

   _____

6. A particle is moving along a straight line. State the condition for which the distance traveled by the particle will be larger than the magnitude of its displacement.

   _____

B. GUIDED PROBLEMS

Note: The solutions to the Guided Problems are given at the end of the Study Guide.

PROBLEM 2.1

The position of a particle that is moving along a horizontal x-axis is given by $x = t^3 - 3t^2 - 5$ in. (x is positive to the right). Calculate the following for the time interval from $t = 0$ to $t = 4$ s: (a) the displacement of the particle; and (b) the total distance traveled by the particle.

a. Comments and Analysis

- The displacement of the particle is the vector drawn from the initial position of the particle to its final position.
- Steps that we will use to calculate the displacement of the particle:

    (1) Find the initial position of the particle.

    (2) Find the final position of the particle.

    (3) Determine the displacement vector.

- If the motion of the particle does not reverse direction, the magnitude of the displacement vector will equal the total distance traveled. However, if the direction of the motion reverses, the total distance traveled will be larger than the magnitude of the displacement vector. Note that the velocity cannot reverse unless the particle comes to rest, that is, unless the velocity is zero.

- Steps that we will use to calculate the total distance traveled:

    (4) Determine if and when the velocity is zero. If the velocity is zero, locate the particle at the time(s) when the velocity is zero. (There may be more than one such time.)

    (5) Sketch the path of the particle and calculate the total distance traveled.

(continued)

b. **Guided Solution**

Part (a)

(1) Find the initial position of the particle.

Using the given value for $x(t)$, calculate the initial position of the particle.

$$x\big|_{t=0} = \underline{\hspace{4cm}}$$

(2) Find the final position of the particle.

Using the given value for $x(t)$, calculate final position of the particle.

$$x\big|_{t=4s} = \underline{\hspace{4cm}}$$

(3) Determine the displacement vector.

On the axes below, place dots at the two points found above.

On the axes below, sketch the displacement vector and label it as $\Delta \mathbf{r}$.

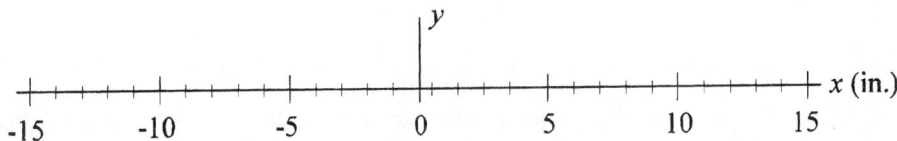

Write the displacement vector using **i** for the unit vector.

$\Delta \mathbf{r} = \underline{\hspace{3cm}}$ **Ans.**

(continued)

5

Part (b)

(4) Determine if and when the velocity is zero.

Using the given value for *x(t)*, derive the expression for the velocity.

v(t) = _____

In the box below, calculate the time(s) when the velocity is zero.

In the box below, find the value of *x* for the time(s) when velocity is zero.

(5) Sketch the path of the particle and calculate the total distance traveled.

On the axes below, place dots at the locations that have been found: the initial position, the final position and the position(s) when the velocity is zero.

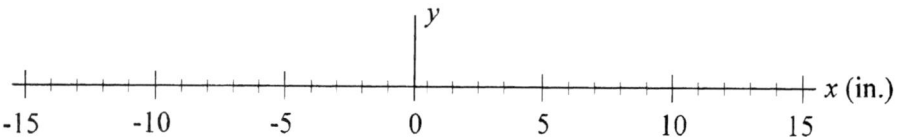

On the axes above, sketch the path of the particle and calculate the total distance traveled.

Total distance traveled = _____ **Ans.**

- -

PROBLEM 2.2

A projectile is fired at point O and follows the parabolic path shown. The coordinates of the projectile as functions of time t in seconds are given by

$$x = 103.9t \text{ in.} \quad \text{and} \quad y = -16.1t^2 + 60t \text{ in.}$$

Determine: (a) the rectangular components of the velocity and acceleration vectors; (b) the magnitude v_0 and direction θ of the initial velocity vector; (c) the maximum height h; and (d) the range R.

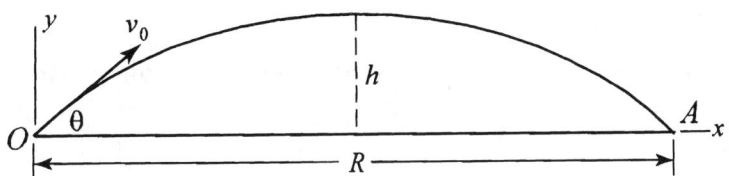

a. Comments and Analysis

- Because the rectangular coordinates as functions of time are given, straight-forward differentiation will determine the velocity and acceleration of the projectile.
- The initial velocity vector, maximum height and range can then be computed from the various kinematic variables.

b. Guided Solution

Part (a) Determine the rectangular components of velocity and acceleration vectors.

Using the given expressions for $x(t)$ and $y(t)$, compute v_x and v_y.

$v_x = $ _____ Ans.

$v_y = $ _____ Ans.

Part (b) Determine the magnitude v_0 and direction θ of the initial velocity vector.

Compute v_x and v_y at $t = 0$.

$v_x\big|_{t=0} = $ _____ $v_y\big|_{t=0} = $ _____

(continued)

In the box below, sketch the initial velocity vector and determine v_0 and θ.

$v_0 = $ _____ **Ans.**

$\theta = $ _____ **Ans.**

Part (c) Determine the maximum height h.

In the box below, calculate the time that the maximum height occurs by solving for the time when $v_y = 0$. (Label that time as t_1.)

$t_1 = $ _____

Calculate h:

$h = $ _____ **Ans.**

Part (d) Determine the range R.

In the box below, calculate the time when the projectile hits the plane at point A by solving for the time when $y = 0$. (Label that time as t_2.)

$t_2 = $ _____

Calculate R:

$R = $ _____ **Ans.**

Lesson 3 Dynamics of Rectilinear Motion: Force-Mass-Acceleration Method

Text Reference: Articles 12.3 and 12.4; Sample Problems 12.5–12.8

A.1 SELF-TEST (*To be done after assigned reading has been completed.*)

1. The method of analysis that uses Newton's laws of motion to relate the forces to the acceleration of a particle is called the _____ method.

2. How would you describe the free-body-diagram of a particle to a student who had never heard of a free-body diagram?

    The free-body diagram of a particle is a sketch of the particle showing

    _____.

3. At a certain instant of time, a particle of mass m is moving with the velocity **v** and acceleration **a**. Write the expression for the inertia vector of the particle at that instant: _____

4. What is the name of the diagram that displays the inertia vector for a particle?

    _____

5. At a certain instant, a particle of mass m is undergoing rectilinear motion along the x-axis with the velocity v_x and the acceleration a_x. The equations of motion for the particle are: $\Sigma F_x =$ _____ ; $\Sigma F_y =$ _____ ; $\Sigma F_z =$ _____

A.2 GENERAL COMMENTS

1. To master the FMA method of kinetic analysis you must be able to correctly construct free-body diagrams. This would be an excellent time to review this technique from your pre-requisite statics course.

(continued)

2. To be able to correctly solve many kinetics problems, it is imperative that you understand the concept of dry friction. (In the following discussion, we let N be the normal force acting between two surfaces.)

Static friction

The maximum static friction force that **can** act between two surfaces at rest is $F_{max} = \mu_s N$, where μ_s is the coefficient of static friction. The static friction force F_s that **does** act between the two surfaces is **always** less than, or equal to, F_{max} $(F_s \leq F_{max})$. The case where $F_s = F_{max}$ is referred to as *impending sliding*.

Kinetic friction

If there is relative motion between two contact surfaces, the kinetic friction force F_k is given by $F_k = \mu_k N$, where μ_k is the coefficient of kinetic friction.

3. If the FMA method of kinetic analysis is used to calculate the acceleration of a particle in an arbitrary position, the velocity and position of the particle can be determined by integration. Integration techniques are discussed in Art. 12.4 in your textbook.

4. Remember that the FMA method of analysis consists of the following steps:
    Step 1: **Draw** the FBD
    Step 2: Use kinematics to **analyze** the acceleration
    Step 3: **Sketch** the MAD
    Step 4: **Write** the equations of motion

B. GUIDED PROBLEMS

PROBLEM 3.1

The block on the 30° inclined plane weighs 10 lb. The coefficients of static and kinetic friction between the block and the plane are $\mu_s = 0.30$ and $\mu_k = 0.15$, respectively.

(a) If the block is released from rest, compute its acceleration. (b) Find the acceleration of the block if it is sliding up the plane with a velocity of 6 ft/s.

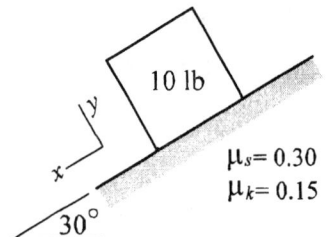

a. Comments and Analysis

- In Part (a), the block is released from rest. Therefore, the first step of the solution will be to determine if the block remains at rest. This part of the solution is essentially a statics problem, with the steps in the analysis being as follows: assume the block remains at rest, compute the required static friction force F_s; and subsequently arrive at one of the following two possibilities: (1) $F_s \leq F_{max}$, in which case the block remains at rest; or (2) $F_s > F_{max}$, in which case the block begins sliding down the plane. If the block slides down the plane, the friction force will equal its kinetic value and will be directed opposite to the direction of motion.

- In Part (b), it is known that the block is sliding up the plane. Therefore, the friction force equals its kinetic value and will be directed down the plane. In this problem, the magnitude of the velocity is irrelevant because only the direction of the velocity is needed.

(continued)

b. Guided Solution

Part (a) Block released from rest.

<u>Assume the block remains at rest</u>

Using the figure at the right, draw the FBD of the block. Label the normal force as N and the static friction force as F_s.

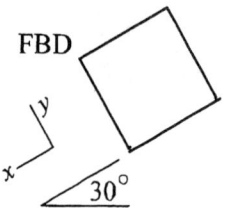

Using the above FBD, compute N and F_s:

$\Sigma F_y = 0$: +↖ _____ = 0   gives $N =$ _____

$\Sigma F_x = 0$: +↗ _____ = 0   gives $F_s =$ _____

Compute the maximum static friction force:

$$F_{max} = \mu_s N = (\quad)(\quad) = \underline{\qquad}$$

Compare the magnitudes of F_s and F_{max}. Is $F_s \leq F_{max}$ ...**YES or NO ?**

Note: If your solution has been correct to this point, you have determined that the block does not remain at rest.

Knowing the block does not remain at rest, using the figure below, draw the FBD and MAD. Label the normal force as N and the kinetic friction force as F_k.

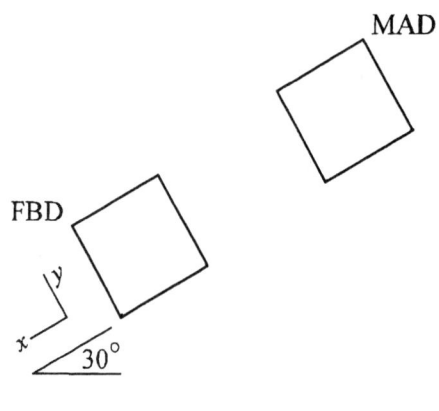

(continued)

Using the above FBD's, complete the solution to determine the acceleration a of the block.

$\Sigma F_y = 0$: $+\nwarrow$ _____ = 0   gives $N =$ _____

$F_k = \mu_k N = ($ $)($ $) =$ _____

$\Sigma F_x = ma$: $+\nearrow$ _____ = _____   gives $a =$ _____ Ans.

Part (b) Block sliding up the plane.

Using the figure below, draw the FBD and MAD knowing that the velocity of the block is directed up the inclined plane. Label the normal force as N and the kinetic friction force as F_k. Be sure to show the correct direction for F_k.

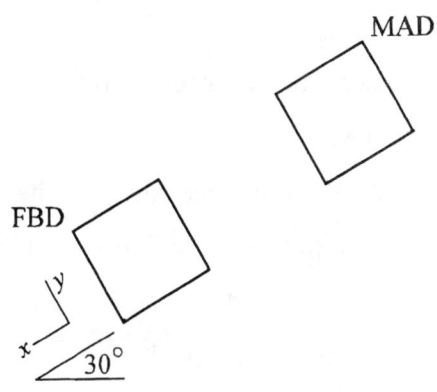

Using the above FBD's, complete the solution to determine the acceleration a of the block.

$\Sigma F_y = 0$: $+\nwarrow$ _____ = 0   gives $N =$ _____

$F_k = \mu_k N = ($ $)($ $) =$ _____

$\Sigma F_x = ma$: $+\nearrow$ _____ = _____   gives $a =$ _____ Ans.

- -

PROBLEM 3.2

At time $t = 0$, the 1.5-kg block is sliding through the position $x = 0$ with the velocity $v_0 = 6$ m/s to the right. At that instant, the force $P = 2t$ N (t is the time in seconds) is applied to slow the block. Determine the acceleration a, velocity v and position x of the block as functions of time. The coefficient of kinetic friction between the block and the plane is 0.3.

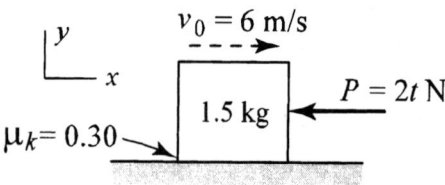

b. Comments and Analysis

- The force-mass-acceleration analysis will determine the acceleration of the block as a function of time. The velocity and position of the block as functions of time can then be found by integration.
- Steps that we will use to determine the acceleration, velocity and position:
    (1) Draw the FBD and MAD.
    (2) Write the equations of motion and calculate the acceleration of the block.
    (3) Integrate the equation of motion to find $v(t)$ and $x(t)$.
    (4) State the initial conditions and evaluate the constants of integration.
    (5) Write the equations for $v(t)$ and $x(t)$.

(continued)

b. Guided Solution

(1) Draw the FBD and MAD.

Using the figure below, draw the FBD and MAD. Label the normal force as N_1 and the friction force as F_1. (Note: Using the subscript is intended to avoid confusion between the normal force N and the N that indicates newtons, the unit of force.) Be sure to show the friction force in the correct direction. Assume that the acceleration a is directed to the right.

(2) Write the equations of motion and calculate the acceleration of the block.

$\Sigma F_y = 0$: $+\uparrow$ _____ gives $N_1 = $ _____

$\Sigma F_x = ma$: $\xrightarrow{+}$ _____

gives $a = $ _____ **Ans.**

(3) Integrate the equation of motion.

Integrate $a(t)$ to find $v(t)$ using C_1 as the constant of integration.

$v(t) = $ _____

Integrate $v(t)$ to find $x(t)$ using C_2 as the constant of integration.

$x(t) = $ _____

(continued)

(4) State the initial conditions and evaluate the constants of integration.

The initial conditions are: (i) at $t = 0$, $v = $ _____ (ii) at $t = 0$, $x = $ _____

In the box below, apply the initial conditions to find the constants of integration.

The values are $C_1 = $ _____ $C_2 = $ _____

(5) Write the equations for $v(t)$ and $x(t)$.

$v(t) = $ _____ **Ans.**

$x(t) = $ _____ **Ans.**

Lesson 4 Curvilinear Motion

Text Reference: Article 12.5; Sample Problems 12.10–12.12

A.1 SELF-TEST (*To be done after assigned reading has been completed.*)

Note: The general forms of the equations of motion written in terms of rectangular coordinates are

$$a_x = f_x(v_x, v_y, v_z, x, y, z, t); \quad a_y = f_y(v_x, v_y, v_z, x, y, z, t)); \quad a_z = f_z(v_x, v_y, v_z, x, y, z, t)$$

1. Stated in words, what does it mean to say that the above general equations of motion are *uncoupled*?

   _____

   _____

2. Write the *uncoupled* forms of the above general equations of motion.

   $a_x =$ $a_y =$ $a_z =$

3. What mathematical methods must usually be used to solve equations of motion that are *coupled*? _____

4. A golf ball is hit from a tee. If air resistance is negligible, are the equations of motion for the ball coupled or uncoupled? _____

A.2 GENERAL COMMENT

Question: For a given problem, how can you tell whether the motion is coupled or uncoupled?

Answer: Draw the FBD, then write and examine the equations of motion.

B. GUIDED PROBLEM

PROBLEM 4.1

The small package of mass m is dropped at A from an airplane that is flying horizontally at an altitude of 400 m with the speed $v_0 = 50$ m/s. Find (a) the time of flight of the package until it hits the ground at B; and (b) the horizontal distance d traveled by the package. Neglect air resistance.

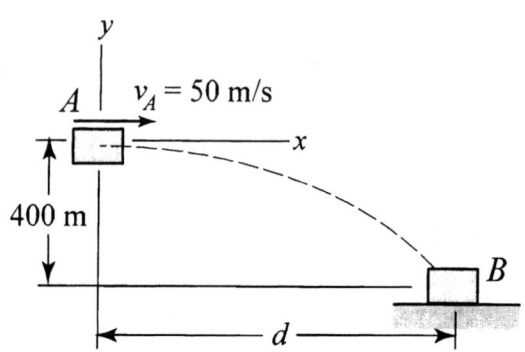

a. Comments and Analysis

- We will use the force-mass-acceleration method to determine the rectangular components of acceleration for the package. Integrating these functions will then yield the components of velocity and position as functions of time (with four constants of integration). The four initial conditions can then be used to evaluate the four constants of integration. Thus, the complete description of the motion will have been found, and all features of the motion can then be determined. (Note: The fact that the problem asks us to find information about the flight of the package does not directly affect the method of solution--the primary challenge is to determine the integration constants.)

- Steps that will be used to determine the time of flight and the horizontal distance d.

  (1) Draw the FBD and MAD.

  (2) Write the equations of motion for the x- and y-directions and determine the acceleration components a_x and a_y.

  (3) Integrate the equations of motion to determine $v_x(t)$, $v_y(t)$, $x(t)$ and $y(t)$.

  (4) State the initial conditions and evaluate the constants of integration.

  (5) Write the final expressions for $v_x(t)$, $v_y(t)$, $x(t)$ and $y(t)$.

  (6) Determine the time of flight for the package.

  (7) Determine the horizontal distance d traveled by the package.

(continued)

b. **Guided Solution**

(1) Draw the FBD and MAD.

Complete the FBD and MAD on the figure below by labeling each vector.

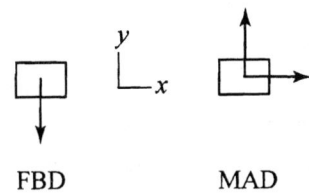

FBD   MAD

(2) Write the equations of motion and determine a_x and a_y.

$\Sigma F_x = ma_x$: $\xrightarrow{+}$ _____ gives $a_x =$ _____

$\Sigma F_y = ma_y$: $+\uparrow$ _____ gives $a_y =$ _____

(3) Integrate the equations of motion to determine $v_x(t)$, $v_y(t)$, $x(t)$ and $y(t)$.

Begin with the above values for a_x and a_y, and identify the constants as C_1 and C_2 (for the x-direction); C_3, and C_4 (for the y-direction).

$a_x =$ _____ $a_y =$ _____

$v_x =$ _____ $v_y =$ _____

$x =$ _____ $y =$ _____

(4) State the initial conditions and evaluate the constants of integration.

Referring to the problem figure, list the four initial conditions on the motion.

(Let $t = 0$ correspond to the time that the package is dropped.)

1. $t = 0$, $x =$ _____   3. $t = 0$, $v_x =$ _____
2. $t = 0$, $y =$ _____   4. $t = 0$, $v_y =$ _____

In the box below, solve for the four constants of integration.

The values are: $C_1 =$ _____; $C_2 =$ _____; $C_3 =$ _____; $C_4 =$ _____

(continued)

(5) Write the final expressions for $v_x(t)$, $v_y(t)$, $x(t)$ and $y(t)$.

$v_x =$ _____ $v_y =$ _____

$x =$ _____ $y =$ _____

(6) Determine the time of flight.

In the box below solve for the time of flight (identify the time as t_1)

$t_1 =$ _____ **Ans.**

(7) Determine d, the horizontal distance traveled.

In the box below solve for d.

$d =$ _____ **Ans.**

Lesson 5 Kinematics: Path (n-t) Coordinates

Text Reference: Articles 13.1 and 13.2; Sample Problems 13.1–13.4

A. SELF-TEST (*To be done after assigned reading has been completed.*)

Note: The following Self-Test refers only to plane motion using path coordinates. (Path coordinates are of limited use in three-dimensional motion.)

1. Define the path coordinate s:

   _____

2. Identify the terms that appear in the following kinematic equations:

   $$\mathbf{v} = v\mathbf{e}_t \qquad \mathbf{a} = a_t\mathbf{e}_t + a_n\mathbf{e}_n \qquad a_n = v^2/\rho \qquad a_t = \dot{v}$$

   (a) $\mathbf{v}$: _____ (b) v: _____

   (c) $\mathbf{a}$: _____ (d) ρ: _____

   (e) a_n: _____ (f) a_t: _____

3. The base vector $\mathbf{e}_n$ is normal to the path and is directed toward

   _____

4. The base vector $\mathbf{e}_t$ is tangent to the path and points in the direction of

   _____

5. Which component of acceleration is due to the change in the magnitude of the velocity? _____

6. Which component of acceleration is due to the change in the direction of the velocity? _____

7. If the motion of a particle is rectilinear, which component of acceleration is zero?

   _____

8. Why is it advantageous to use the equation $a_t\, ds = v\, dv$ in some situations?

   _____

B. GUIDED PROBLEMS

PROBLEM 5.1

A particle moves along a circular path of radius $R = 15$ m. The path length s, measured from a fixed reference point on the path, varies as $s(t) = 4t^3 - 10t$ m, where t is the time in seconds. Calculate the following at $t = 3$ s: (a) the magnitude of the velocity vector **v**; (b) the magnitude of the acceleration vector **a**; and (c) the angular velocity and angular acceleration of the radial line.

a. Comments and Analysis

- For motion along a circular path of radius R, it is convenient to use path (n-t) coordinates, with the kinematic equations being:

$$s = R\theta \quad v = \frac{ds}{dt} = R\dot\theta \quad a_t = \frac{dv}{dt} = R\ddot\theta \quad a_n = \frac{v^2}{R} \quad a_t\,ds = v\,dv$$

b. Guided Solution

Part (a)

Using the given expression for $s(t)$, calculate the magnitude of the velocity vector $v(t)$:

$v(t) = $ _____

Evaluate $(v)_{t=3\,s} = $ _____ **Ans.**

Part (b)

Using the results of Part (a), calculate the normal component of acceleration at $t = 3$ s:

$(a_n)_{t=3\,s} = $ _____

Using the results of Part (a), determine the tangential component of acceleration as a function of time: $a_t = $ _____. **Evaluate** $(a_t)_{t=3\,s} = $ _____.

Using the above results, compute the magnitude of the acceleration at $t = 3$ s:

$(a)_{t=3\,s} = $ _____ **Ans.**

Part (c)

Using the above results, calculate the following:

Angular velocity: $(\dot\theta)_{t=3\,s} = $ _____ **Ans.**

Angular acceleration: $(\ddot\theta)_{t=3\,s} = $ _____ **Ans.**

PROBLEM 5.2

A particle is moving along the curved path shown. As the particle passes point O, it speed is 10 ft/s and the magnitude of its total acceleration is 17 ft/s².

The speed v of the particle changes according to $(-ks + 8)$ ft/s², where s is the distance measured along the path from point O and k is a constant. When $s = 6$ ft, $v = -4$ ft/s. Determine the radius of curvature of the path at point O.

b. Comments and Analysis

- When interested in the radius of curvature of a path, it is convenient to use path (n-t) coordinates, with the kinematic equations being:

$$v = \frac{ds}{dt} \qquad a_t = \frac{dv}{dt} \qquad a_n = \frac{v^2}{\rho} \qquad a_t\, ds = v\, dv$$

- Steps that we will use to determine the radius of curvature of the path at point O:

    (1) Determine v^2 as a function of s.   (3) Calculate a_n at point O.
    (2) Calculate a_t at point O.   (4) Calculate ρ at point O.

b. Guided Solution

(1) Determine v as a function of s.

Note: The problem statement reads, "the speed v of the particle changes according to $(-ks + 8)$ ft/s²." You must recall that the tangential component of acceleration equals the time rate of change of the speed, that is $a_t = (-ks + 8)$ ft/s².

(a) Integrate $v\, dv = a_t\, ds$ to obtain $v(s)$. Let the constant of integration be C_1.

$$\int v\, dv = \int a_t\, ds = \int (-ks + 8)\, ds$$

$$\frac{v^2}{2} =$$

(b) Write the two known conditions that relate v to s.

(i) _____ and (ii) _____

(continued)

(c) Evaluate the two constants k and C_1.

   (i) gives:

   (ii) gives:

(d) Write the final expression for $v^2(s)$. $v^2(s) =$ _____

(2) Calculate a_t at point O.

The result is $a_t =$ _____

(3) Knowing a_t and the given value for the magnitude of the total acceleration at O, calculate a_n at point O.

The result is $a_n =$ _____

(4) Calculate ρ at point O.

The result is $\rho =$ _____ **Ans.**

PROBLEM 5.3

As the pin P moves along the circular slot, v_x (the horizontal component of its velocity) is constant. When P is in the position shown, the magnitude of its acceleration is 36 m/s^2. Determine the magnitude of the velocity of P in this position

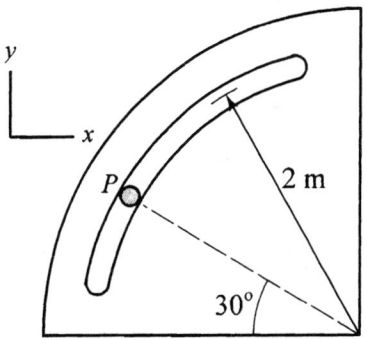

c. Comments and Analysis

- The solution to this problem combines path coordinates (n-t) and rectangular (x-y) coordinates—path coordinates to describe the particle motion on a circular path, and rectangular coordinates because the problem states that v_x is constant.

b. Guided Solution

(a) The problem states that v_x is constant. Therefore, what is known about the direction of the acceleration of the pin? _____

(b) In the order that they are listed, sketch the following on the figure below
  (i) the acceleration vector for the pin (include its magnitude which is given).
  (ii) the normal and tangential components of the acceleration of the pin.

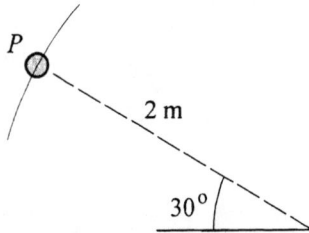

Question: Why is the acceleration in the figure directed downward instead of upward?
Answer: _____

(c) From the geometry of the above figure, calculate the normal component of acceleration. $a_n =$ _____

(d) Calculate the magnitude of the velocity of P for the position shown.

$v =$ _____ **Ans.**

Lesson 6 Kinematics: Polar and Cylindrical Coordinates

Text Reference: Articles 13.3; Sample Problems 13.5–13.8

A. SELF-TEST *(To be done after assigned reading has been completed.)*

1. The base vector $\mathbf{e}_R$ in the polar coordinate system is directed away from the origin of the *xy*-coordinate system and is directed along the _____ line.

2. The base vector $\mathbf{e}_\theta$ in the polar coordinate system is perpendicular to (a) _____ and points in the direction of increasing (b) _____.

3. The velocity vector described in polar coordinates is $\mathbf{v} = v_R \mathbf{e}_R + v_\theta \mathbf{e}_\theta$, where
   (a) $v_R =$ and (b) $v_\theta =$

4. The acceleration vector described in polar coordinates is $\mathbf{a} = a_R \mathbf{e}_R + a_\theta \mathbf{e}_\theta$, where
   (a) $a_R =$ and (b) $a_\theta =$

5. For the special case where the path of the particle is a circle, the polar coordinate R is a constant equal to _____.

6. In cylindrical coordinates, the base vectors are $\mathbf{e}_R$, $\mathbf{e}_\theta$, and $\mathbf{e}_z$. How is $\mathbf{e}_z$ related to the base vectors ($\mathbf{i}, \mathbf{j}, \mathbf{k}$) of the Cartesian frame? $\mathbf{e}_z =$

B. GUIDED PROBLEM

PROBLEM 6.1

The plane motion of the particle P described in polar coordinates is $R = 0.4t^2$ m, $\theta = t^3/8$ rad, where t is the time in seconds. When $t = 2$ s, determine the magnitude of (a) the velocity vector; and (b) the acceleration vector. Show each vector on a sketch.

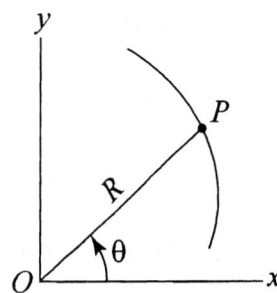

a. Comments and Analysis

- The solution will involve evaluating the time derivates of the given expressions for R and θ at $t = 2$ s. The appropriate values will then be substituted into the relevant kinematic equations for particle motion in polar coordinates:

$$v_R = \dot{R} \quad v_\theta = R\dot{\theta} \quad a_R = \ddot{R} - R\dot{\theta}^2 \quad a_\theta = R\ddot{\theta} + 2\dot{R}\dot{\theta}$$

b. Guided Solution

Computation of derivatives (Be sure to include the units for each term.)

(a) Complete the following table that gives the first and second time derivates of the polar coordinates. Be sure to include the units for each term.

$R =$ $\theta =$

$\dot{R} =$ $\dot{\theta} =$

$\ddot{R} =$ $\ddot{\theta} =$

(b) Complete the following table that gives the values of the time derivatives at $t = 2$ s.

$R\big|_{t=2\,\text{s}} =$ $\theta\big|_{t=2\,\text{s}} =$

$\dot{R}\big|_{t=2\,\text{s}} =$ $\dot{\theta}\big|_{t=2\,\text{s}} =$

$\ddot{R}\big|_{t=2\,\text{s}} =$ $\ddot{\theta}\big|_{t=2\,\text{s}} =$

(continued)

Part (a)

Using the tabulated results, evaluate the polar components of the velocity at $t = 2$ s.

$v_R = \dot{R} =$ _____ $v_\theta = R\dot{\theta} =$ _____

Compute the magnitude of the velocity vector.

$v =$ The result is $v =$ _____ **Ans.**

On Fig. (a) below, sketch the velocity vector approximately to scale. Indicate and calculate the angle between the velocity vector and the radial line.

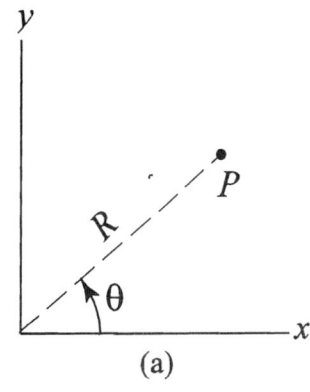 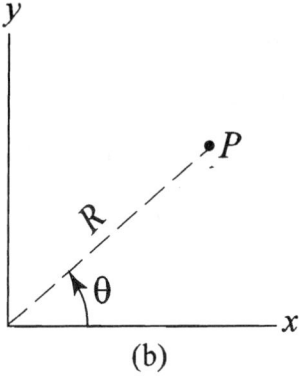

            (a)   (b)

Part (b)

Using the tabulated results, evaluate the polar components of the acceleration at $t = 2$ s.

$a_R = \ddot{R} - R\dot{\theta}^2 =$ _____

$a_\theta = R\ddot{\theta} + 2\dot{R}\dot{\theta} =$ _____

Compute the magnitude of the acceleration vector.

$a =$ The result is $a =$ _____ **Ans.**

On Fig. (b) above, sketch the acceleration vector approximately to scale. Indicate and calculate the angle between the acceleration vector and the radial line.

c. Concluding Remark

The problem statement described the motion in terms of polar coordinates. However, having solved for the velocity vector, we know the normal and tangential directions—the velocity vector is tangent to the path. Therefore, a_n and a_t could now be found by geometry.

Lesson 7 Kinetics: Force-Mass-Acceleration Method, Curvilinear Coordinates

Text Reference: Article 13.4; Sample Problems 13.9–13.12

A.1 SELF-TEST (*To be done after assigned reading has been completed.*)

1. List the four basic steps in the force-mass-acceleration (FMA) method of kinetic analysis of a particle :

   (1) Draw the _____

   (2) Perform kinematic analysis of the _____

   (3) Draw the _____

   (4) Derive the _____

2. Of the four basic steps in the FMA method that are numbered in Question 1, which step is significantly different if curvilinear coordinates are used instead of Cartesian coordinates? Step _____

A.2 GENERAL COMMENT

There is no new material introduced in the assigned textbook article. The equations presented are the equations that result from writing the equation $\Sigma \mathbf{F} = m\mathbf{a}$, where the acceleration $\mathbf{a}$ is expressed in terms of curvilinear coordinates.

B. GUIDED PROBLEMS

PROBLEM 7.1

The 0.75-kg slider is fired upward from A along the frictionless curved rod that lies in the vertical plane. As the slider passes point B with a speed of 3 m/s, determine the magnitude R of the force that the rod exerts on the slider; and the rate at which the speed of the slider is changing. (Is it increasing or decreasing?)

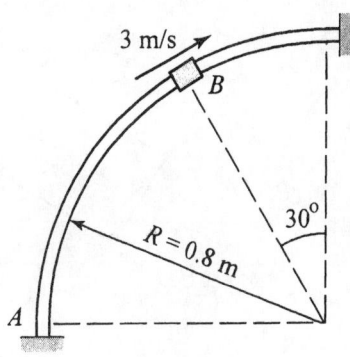

(continued)

a. **Comments and Analysis**

- The force-mass-acceleration method is well-suited for determining the required quantities at the instant the slider is at point B.
- Since the rod is frictionless, the force exerted by the rod is normal to the rod.
- Because the path of the slider is circular, we will use normal and tangential (n-t) components to describe the kinematics of motion.
- Recall that "rate of change of speed" of the slider is measured by the magnitude of the tangential component its acceleration.
- Steps that we will use to determine the R and a_t:

    (1) Determine the n-t components of the inertia vector.

    (2) Draw the FBD and MAD.

    (3) Count the number of unknowns and independent equations of motion.

    (4) Write the equations of motion and solve for R and a_t.

b. **Guided Solution**

(1) Determine the n-t components of the inertia vector.

   Evaluate the following:

   Weight $W = $ _____

   $a_n = $ _____

   $ma_n = $ _____

   $ma_t = $ _____ (Recall that m is known, but a_t is an unknown quantity.)

(2) Draw the FBD and MAD on the figures below. (**Note**: the sense of R is arbitrary.)

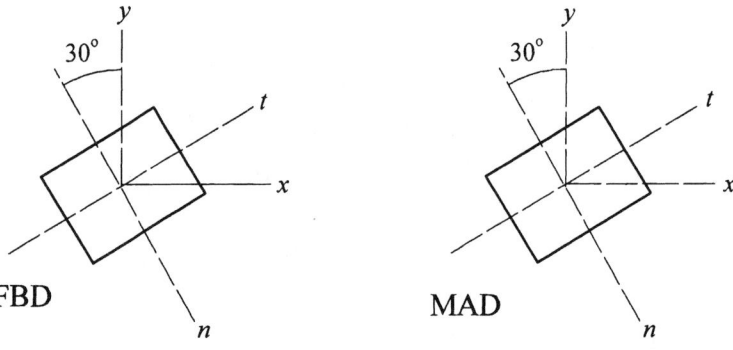

(continued)

(3) Count the number of unknowns and the number of independent equations of motion.

The total number of unknowns is ____ and the number of independent equations of motion is ____.

(4) Write the equations of motion and solve for R and a_t.

$\Sigma F_n = ma_n$: ↙+ _____ gives $R =$ _____ Ans.

$\Sigma F_t = ma_t$: +↗ _____ gives $a_t =$ _____ Ans.

Is the speed increasing or decreasing? _____

PROBLEM 7.2

The 5000-kg rocket was fired vertically. At a certain time, the radar tracking station at O recorded the following data:

$$\theta = 60° \qquad R = 6500 \text{ m}$$
$$\dot{R} = 56.3 \text{ m/s} \qquad \dot{\theta} = 0.02 \text{ rad/s} \qquad \ddot{\theta} = 0.003 \text{ rad/s}^2$$

Determine the magnitude of the thrust P at this time. Neglect air resistance and use $g = 9.81 \text{ m/s}^2$.

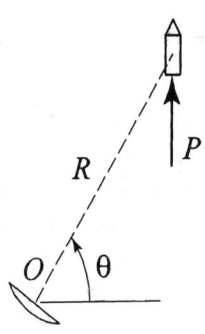

b. Comments and Analysis

- Note that the data have been recorded using polar coordinates. Recall that the magnitudes of the acceleration components in polar coordinates are

$$a_R = \ddot{R} - R\dot{\theta}^2 \quad \text{and} \quad a_\theta = R\ddot{\theta} + 2\dot{R}\dot{\theta}$$

- Steps that we will use in the analysis:

   (1) Draw the FBD to determine the direction of the acceleration vector.

   (2) Using the recorded data, compute the magnitude of the acceleration vector.

   (3) Compute the thrust P by analyzing the FBD and MAD.

(continued)

b. **Guided Solution**

(1) Draw the FBD for the rocket on the figure at the right. Inspection of this FBD reveals that the direction of the acceleration is _____ .

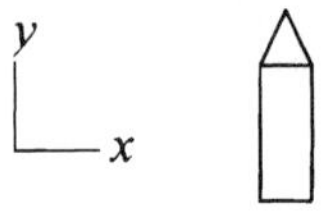

(2) On the figure at the right, label the acceleration vector **a** and its a_R- and a_θ-components. Using the given data, calculate a_θ:

$a_\theta = $ _____

which gives $a_\theta = $ _____ m/s². Using this value of a_θ, use the geometry of the figure to compute the acceleration a:

$a = $ _____

which gives $a = $ _____ m/s².

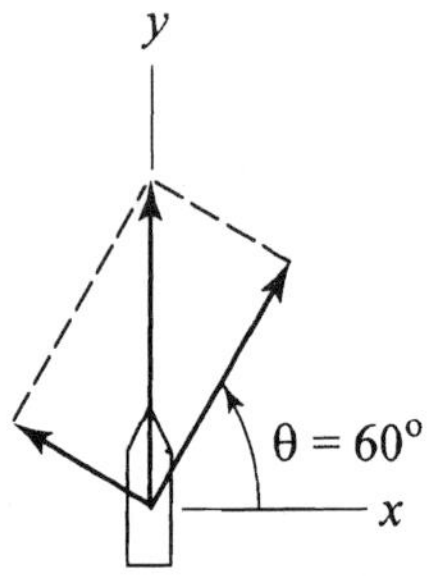

Acceleration

(continued)

(3) Draw the FBD and MAD for the rocket.

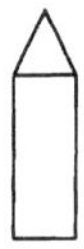

FBD

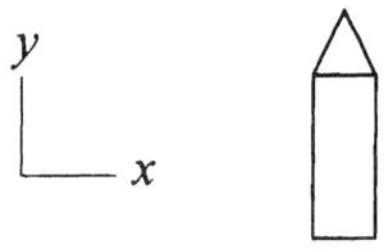

MAD

In the box below, write the equation of motion and solve for the thrust P.

$P = $ _____ **Ans.**

> **Lesson 8 Work of a Force; Principle of Work and Kinetic Energy**
>
> *Text Reference*: Articles 14.1–14.3; Sample Problems 14.1–14.3

A. SELF-TEST (*To be done after assigned reading has been completed.*)

1. The point of application of a force **F** undergoes the differential displacement *d**r***. Write the defining equation for the differential work done by **F**: $dU =$

2. A force **F** moving along a path is expressed in terms of two components: F_n, the component perpendicular to the path, and F_t, the component tangent to the path. Which of these components is the *working component* of **F**? _____

3. The point of application of a force **F** undergoes the differential displacement *d**r***. What is meant by the *work-absorbing component* of *d**r***?
   _____

4. (a) The work done by a variable force as it moves between two points depends upon the path taken by the point of application of the force. (T or F) _____
   (b) The work done by a constant force as it moves between two points depends upon the path taken by the point of application of the force. (T or F) _____

5. A 180-lb man skis 100 ft down a 45° slope. Calculate the work done by the man's weight.
   $U_{1-2} =$

6. A spring with constant 80 N/m is deformed from a compression of 600 mm to an elongation of 800 mm. Calculate the work done *by* the spring:
   $U_{1-2} =$

7. A particle of mass 0.8 kg is traveling at a speed of 4 m/s. Calculate the kinetic energy of the particle:
   $T =$

(continued)

8. A constant 10-N force acting in the positive *x*-direction moves 5 m in the positive *x*-direction. Why is it not meaningful to state that the work done by the force is 50 N·m in the *x*-direction?

   _____

9. State the work-energy principle for a particle in words.

   _____

   _____

10. A free-body diagram shows all the forces that act on a particle. An *active-force diagram* shows only _____ .

B. GUIDED PROBLEMS

PROBLEM 8.1

The 3-kg collar is attached to the linear spring with stiffness $k = 24$ N/m and unstretched length $L_0 = 1.75$ m. The collar starts from rest at A and slides up the frictionless rod under the action of the constant 40-N force. Determine the velocity of the collar as it passes B.

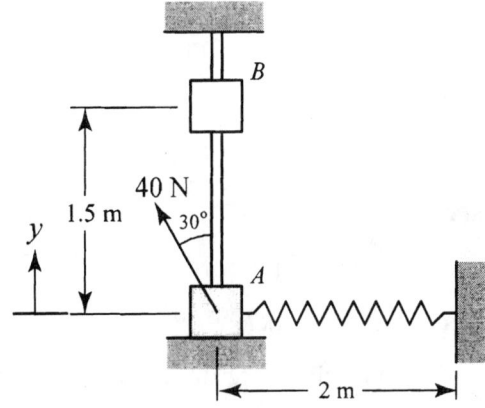

a. Comments and Analysis

- Work-energy is a convenient method of analysis because the problem involves a change in speed that occurs over a change in distance.
- The steps that we will use in determine the velocity at B are:
    (1) Determine which forces do work on the collar as it moves from the initial position A to the final position B. Calculate U_{A-B}, the total work done on the collar.
    (2) Write the expressions for T_A and T_B, the kinetic energy of the collar in positions A and B, respectively.
    (3) Substitute into the work-energy relation, $U_{A-B} = T_B - T_A$, and solve for the velocity of the collar as it passes B.

(continued)

b. Guided Solution

(1) Determine which forces do work on the collar and calculate U_{A-B}.

The figure at the right shows the FBD of the collar when it is in an arbitrary position. The four forces include its weight W, the spring force F_s, the applied 40-N force, and the normal force N exerted by the vertical rod (no friction force because the rod is frictionless). All the forces except N do work on the collar. Why does N not do work? _____

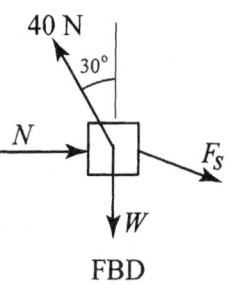

FBD

Compute the work done by each of the three forces.

Weight W: $U_{A-B} = -Wy =$

40-N applied force: $U_{A-B} = Fd =$

Spring force F_s: $L_0 = 1.75$ m $L_A =$ ___ m $L_B = \sqrt{()^2 + ()^2} =$ ___ m

$\delta_A =$ _____ $\delta_B =$ _____

Substituting into $U_{A-B} = -\dfrac{1}{2}k(\delta_B^2 - \delta_A^2)$ gives:

$U_{A-B} =$

(2) Write the expressions for the kinetic energy of the collar in positions A and B (let v_B refer to the velocity of the collar in position B).

$T_A =$ _____   and $T_B =$ _____

(3) In the box below, substitute into the work-energy relation and solve for v_B.

$v_B =$ _____ m/s **Ans.**

PROBLEM 8.2

In the position shown, the velocity of the 16.1-lb block A is $v_1 = 36$ ft/s up the inclined plane. The coefficient of kinetic friction between the block and the plane is $\mu_k = 0.5$ and the spring constant is $k = 100$ lb/ft. Neglect the weight of the striker plate B and assume the spring is initially stress-free. Determine the deflection Δ of the spring when the block comes to rest.

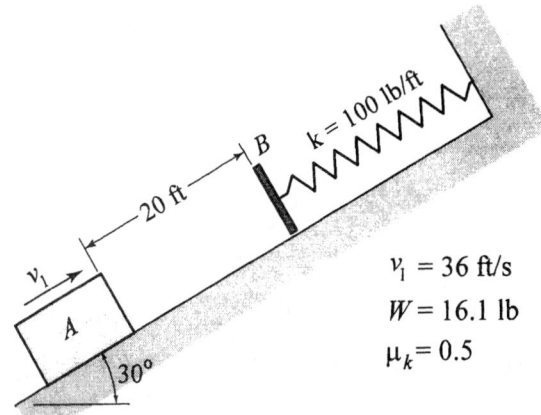

$v_1 = 36$ ft/s
$W = 16.1$ lb
$\mu_k = 0.5$

b. Comments and Analysis

- Work-energy is a convenient method of analysis because the problem involves a change in speed that occurs over a change in distance.
- The subscript 1 refers to the initial position shown, and we will use the subscript 2 to refer to the final position where the block comes to rest. We will assume the spring deflection Δ is measured in feet.
- The steps that we will use in determine the deflection Δ are:

    (1) Determine the forces that do work as the block moves from the initial position 1 to the final position 2. Calculate U_{1-2}, the total work done.

    (2) Write the expressions for T_1 and T_2, the kinetic energy of the block in positions 1 and 2, respectively.

    (3) Substitute into the work-energy relation, $U_{1-2} = T_2 - T_1$, and solve for the spring deflection Δ.

- Note that no energy is lost when the block hits the striker plate B because the weight of the striker plate is being neglected. The impact of two bodies with non-negligible weights is the topic of a later lesson.

(continued)

b. **Guided Solution**

(1) Determine which forces do work on the block and calculate U_{1-2}.

The figure at the right shows the FBD of the block as it slides up the plane. The four forces include its weight W, the normal force N, the kinetic friction force F_k, and the spring force F_s. (The spring force is shown as dotted because that force is zero until the block has traveled more than 20 ft up the plane.) All the forces except N do work on the collar. Why does N not do work? _____

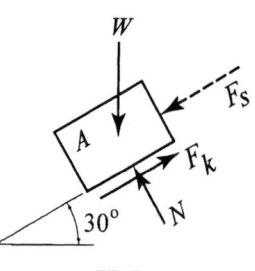

FBD

Compute the work done by each of the three forces between positions 1 and 2.

Weight W: $U_{1-2} = -Wh =$

(h is vertical distance moved upward between positions 1 and 2)

Kinetic friction force F_k:

Referring to the FBD: $N =$

$$F_k = \mu_k N =$$

Therefore, $U_{1-2} = -F_k d =$

Spring force F_s:

$\delta_1 =$ _____ $\delta_2 =$ _____   Substituting into $U_{1-2} = -\frac{1}{2}k(\delta_2^2 - \delta_1^2)$ gives:

$U_{1-2} =$

---

(2) Write the expressions for the kinetic energy of the block in positions 1 and 2.

$T_1 =$ _____   and $T_2 =$ _____

(3) In the box below, substitute into the work-energy relation and solve for Δ.

$\Delta =$ _____ ft **Ans.**

Lesson 9 Conservative Forces and Conservation of Mechanical Energy

Text Reference: Article 14.4; Sample Problems 14.4 and 14.5

A. SELF-TEST *(To be done after assigned reading has been completed.)*

1. A force is said to be *conservative* if its work depends only upon
   _____.

2. The capacity of a force to do work is called its _____.

3. Complete the following:
   the work of a conservative force = the decrease in its _____

4. The *total mechanical energy* is the sum of the _____ and _____ energies.

5. Kinetic friction is a nonconservative force because its work is not independent of _____.

6. The *principle of conservation of mechanical energy* is valid only for conservative systems. (T or F) ____

7. Write the expression for the potential energy of a spring with constant k and deformation δ: $V_e =$

8. Write the expression for the potential energy of a weight W that is located at a distance h above the datum: $V_g =$

B. GUIDED PROBLEM

PROBLEM 9.1

The vertical spring with constant $k = 300$ N/m is attached to the 0.8-kg block. The block is initially at rest in the equilibrium position shown. The block is then pulled down a distance of 400 mm and released. Determine the velocity of the block as it passes through the equilibrium position. Solve using the principle of conservation of mechanical energy.

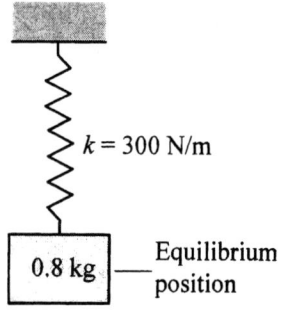

a. Comments and Analysis

- Both the weight of the block and the spring force are conservative so we are justified in using the principle of conservation of energy. (A solution using the work-energy principle would be as convenient.)
- We will use the subscript 1 to refer to the initial position of release, and the subscript 2 to refer to the final position (the equilibrium position).
- The steps that we will use to determine v_2, the velocity of the block in position 2, are:
    (1) Choose the datum for the potential energy V_g. Calculate the gravitational potential energy in positions 1 and 2.
    (2) Calculate the elastic potential energy V_e in positions 1 and 2.
    (3) Write the expressions for T_1 and T_2, the kinetic energy of the block in positions 1 and 2, respectively.
    (4) Apply the principle of conservation of mechanical energy and solve for v_2.

(continued)

b. **Guided Solution**

(1) The figure at the right shows that we have chosen position 1 to be the position from which the block is released, and position 2 to be the equilibrium position. The figure also shows that we have taken the datum for V_g to be the location of the block in position 1. Calculate the gravitational potential energies: $(V_g)_1 = Wy_1 = $ _____ and $(V_g)_2 = Wy_2 = $ _____.

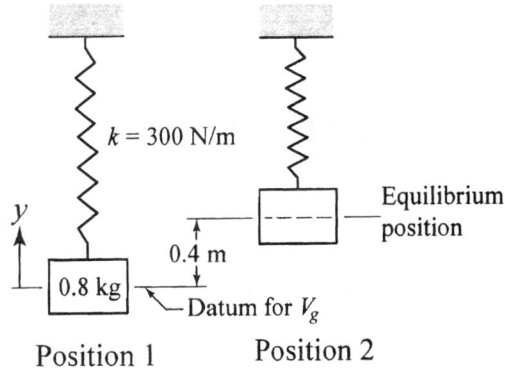

(2) *Elastic potential energy V_e:*

When the block is in the equilibrium position (position 2), the spring force is equal to its weight. Determine the spring elongations:

$$\delta_2 = \frac{W}{k} = \underline{\hspace{3cm}}$$

$$\delta_1 = \delta_2 + 0.4 \text{ m} = \underline{\hspace{3cm}}$$

Calculate the elastic potential energies:

$$(V_e)_1 = \frac{1}{2}k\delta_1^2 = \underline{\hspace{4cm}}$$

$$(V_e)_2 = \frac{1}{2}k\delta_2^2 = \underline{\hspace{4cm}}$$

(3) Write the expressions for the kinetic energies T_1 and T_2 (let v_2 be the velocity in position 2):

$T_1 = $ _____ and $T_2 = $ _____

(4) Apply the principle of conservation of mechanical energy and solve for v_2.

Substitute the above terms into $(V_g)_1 + (V_e)_1 + T_1 = (V_g)_2 + (V_e)_2 + T_2$:

$$\underline{\hspace{5cm}} = \underline{\hspace{5cm}}$$

which gives: $v_2 = $ _____ **Ans.**

Lesson 10 Power and Efficiency

Text Reference: Article 14.5; Sample Problems 14.6 and 14.7

A. SELF-TEST (*To be done after assigned reading has been completed.*)

1. Define power (in words): _____

2. Write the defining equation for the power P in terms of the work U: $P =$

3. Write the equation for obtaining the power P of a force **F** that acts on a particle moving with velocity **v**: $P =$

4. A 10 lb force acts on a particle that is moving along the *x*-axis at a speed of 4 ft/s. Why is it incorrect to state that the power of the particle is $P = Fv = 10(4) = 40$ lb·ft/s in the *x*-direction? _____

5. The rate at which energy is supplied to a machine is called _____ power.

6. The rate at which a machine does work is called _____ power.

7. Write the equation that defines the efficiency η of a machine:

   $\eta =$

B. GUIDED PROBLEM

PROBLEM 10.1

The hydrodynamic force F that opposes the motion of a boat is proportional to the square of the speed v of the boat, that is, $F = cv^2$, where c is a constant. The boat's motor delivers the power $P_1 = 13.4$ kW to move the boat at a constant speed of $v_1 = 4.5$ m/s. What power P_2 is required for a constant speed of $v_2 = 6$ m/s? Compare the percentage increase in speed with the percentage increase in required power.

a. Comments and analysis

Question 1: Power P is defined as the dot product of the force vector **F** and velocity vector **v**: $P = \mathbf{F} \cdot \mathbf{v}$. In this problem, it is permissible to use $P = Fv$. Why?

Answer: _____

Question 2: The force supplied by the motor is equal to the hydrodynamic force. Why?

Answer: _____

b. Guided Solution

(a) In the box below, form the ratio P_2/P_1 knowing that $P = cv^3$. Using the given data, determine P_2.

$$P_2 = \underline{\hspace{2cm}} \text{ Ans.}$$

(b) Calculate the required percentage increases:

   (i) Percent increase in speed =

   _____% Ans.

   (ii) Percent increase in required power =

   _____% Ans.

Lesson 11 Principle of Impulse and Momentum

Text Reference: Article 14.6; Sample Problems 14.8 and 14.9

A. SELF-TEST (*To be done after assigned reading has been completed.*)

1. Write the defining equation for the *impulse* of a force **F** for the time interval t_1 to t_2:

    $\mathbf{L}_{1-2} =$

2. The impulse of a force can be zero, even if the force is not zero. (T or F) ____

3. A force can have an impulse even if it does not do work. (T or F) ____

4. Write the defining equation for the *momentum* of a particle of mass m traveling with the velocity **v**: $\mathbf{p} =$

5. How is the resultant force $\Sigma \mathbf{F}$ acting on a particle related to its linear momentum **p**?

    $\Sigma \mathbf{F} =$

6. Write the impulse-momentum principle:

    $\mathbf{L}_{1-2} =$

7. The momentum of a particle is conserved if $\mathbf{L}_{1-2} =$

8. A component of momentum can be conserved, even if the total momentum is not conserved. (T or F) ____

9. Consider the four terms in the two equations: $U_{1-2} = \Delta T$ and $\mathbf{L}_{1-2} = \Delta \mathbf{p}$. Identify the term that is associated with each of the following:

    (a) a force and a change in time:

    (b) a force and a change in position:

    (c) the mass of a particle and the change in its velocity vector:

    (d) the mass of a particle and the change in its speed:

B. GUIDED PROBLEM

PROBLEM 11.1

The 200-kg crate on rollers is being pulled up the inclined plane by the 700-N constant force, and the force $P(t)$ that varies with time as shown. (Both of these forces are parallel to the plane). If the crate starts from rest at $t = 0$, determine its velocity when $t = 7$ s. Assume the masses of the rollers are negligible.

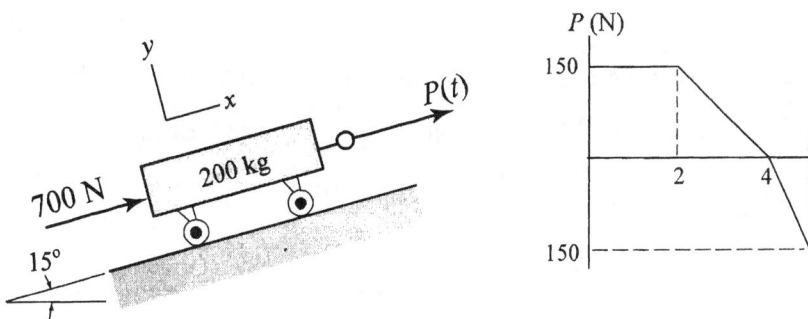

a. Comments and Analysis

- Either FMA or impulse-momentum could be used to solve a problem such as this that involves a change in velocity that occurs over a time interval. However the FMA method would be cumbersome because the acceleration is different for each time period: 0–2 s, 2–4 s, etc. Therefore, we choose to solve this problem using the impulse-momentum principle.

- We will use the subscript 1 to refer to the initial time ($t_1 = 0$) and the subscript 2 to refer to the final time ($t_2 = 7$s).

- The steps that we will use to determine v_2, the velocity of the block when $t = t_2$ are:

    (1) Draw the FBD of the crate at an arbitrary time. (The FBD is necessary to ensure that all forces acting on the crate are considered when computing the impulse.)

    (2) Calculate the impulse in the x-direction $(L_{1\text{-}2})_x$ for each force.

    (3) Write the expressions for momenta $(p_x)_1$ and $(p_x)_2$ for the crate.

    (4) Write the impulse-momentum equation for the x-direction and solve for v_2.

(continued)

b. Guided Solution

(1) Draw the FBD of the crate using the figure at the right. (Let R represent the resultant normal force acting on the crate.)

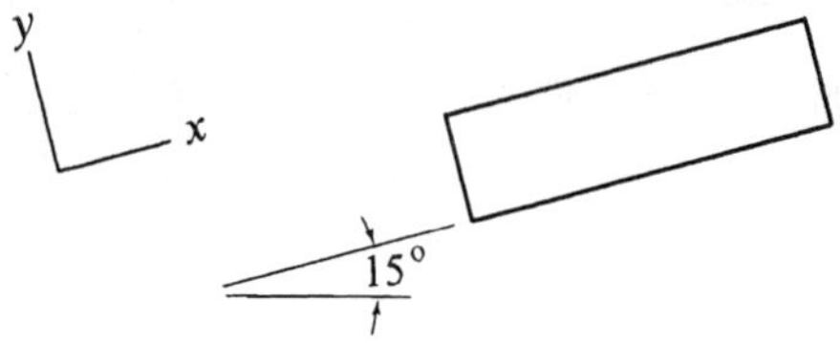

   Note: Since the masses of the rollers are negligible, there is no friction force.

(2) Calculate the impulse in the x-direction for each of the forces for the time period t_1 to t_2. (The impulse-momentum analysis in the y-direction would yield $R = W\cos 15°$, a result that is also readily obtainable by summing forces in the y-direction.)

   Normal force R: $(L_{1\text{-}2})_x =$ _____   *700-N force:* $(L_{1\text{-}2})_x =$ _____

   Weight: $(L_{1\text{-}2})_x =$ _____

   $P(t)$: $(L_{1\text{-}2})_x =$ _____

(3) Write the expressions for momenta $(p_x)_1$ and $(p_x)_2$ for the crate.

   $(p_x)_1 =$ _____ $(p_x)_2 =$ _____

(4) In the box below, write the impulse-momentum equation for the x-direction and solve for v_2.

$(L_{1\text{-}2}) = (p_x)_2 - (p_x)_1$

$\xrightarrow{+}$

$v_2 =$ _____ (up or down the plane?) **Ans.**

Lesson 12 Principle of Angular Impulse and Momentum

Text Reference: Article 14.7; Sample Problems 14.10 and 14.11

A. SELF-TEST (*To be done after assigned reading has been completed.*)

1. Write the defining equation for the *angular impulse* of a force **F** about point *A* for the time interval t_1 to t_2 in terms of $\mathbf{M}_A$, the moment of **F** about *A*:

    $(\mathbf{A}_A)_{1-2} =$

2. What are the dimensions of angular impulse? _____

3. Identify each of the terms in the defining equation for $\mathbf{h}_A$, the *angular momentum* of a particle about point *A*: $\mathbf{h}_A = \mathbf{r} \times m\mathbf{v}$

    (a) **r** : _____

    (b) *m*: _____

    (c) **v**: _____

4. The equation $\mathbf{M}_A = \dot{\mathbf{h}}_A$ is valid for every point. (T or F) ____

5. Write the *principle of angular impulse and angular momentum*:

    $(\mathbf{A}_A)_{1-2} =$ \hspace{3cm} (*A*: fixed point)

6. If $(\mathbf{A}_A)_{1-2} = \mathbf{0}$, the angular momentum about point *A* is conserved. What restriction is placed upon the choice of point *A*? _____

7. A component of angular momentum can be conserved, even if the total angular momentum is not conserved. (T or F) ____

47

B. GUIDED PROBLEM

PROBLEM 12.1

The small ball A of known mass m is attached to one end of a light cord that passes through the hollow support at O. Initially the cord is rotating about the z-axis at the angular speed $\omega_1 = 10$ rad/s with the ball moving around a circular path of radius $r_1 = 1.2$ m. The cord is then drawn slowly through the support until the final radius of the path is $r_2 = 0.6$ m. Determine ω_2, the final angular velocity of the cord.

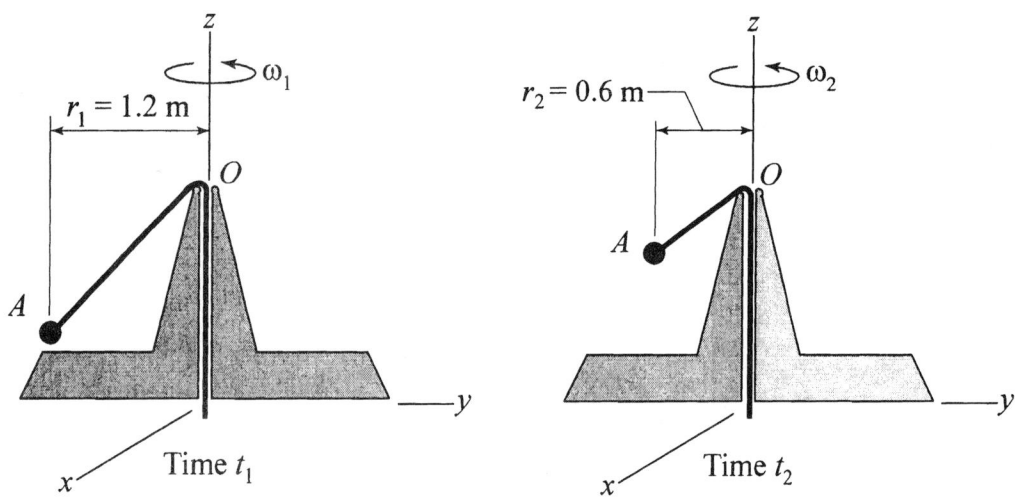

a. Comments and Analysis

- We will use the angular impulse-momentum principle to analyze this problem.
- The steps that we will use to determine ω_2 are:

    (1) Draw the FBD of the ball A at an arbitrary time. Show that the angular impulse about the z-axis is zero throughout the motion.

    (2) Draw the momentum diagram at time t_1. Calculate $(h_z)_1$, the initial angular momentum about the z-axis.

    (3) Draw the momentum diagram at time t_2. Calculate $(h_z)_2$, the final angular momentum about the z-axis.

    (4) Use the angular impulse-momentum principle to determine ω_2.

(continued)

b. **Guided Solution**

(1) **Draw** the FBD of ball A at an arbitrary time in Fig. (a).

   Question: Why is the angular impulse about the z-axis zero throughout the motion?

   Answer: $(A_z)_{1\text{-}2} = 0$ because _____

TOP VIEW

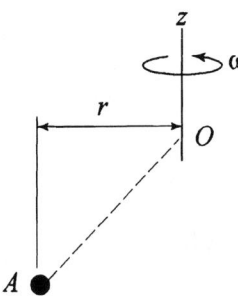
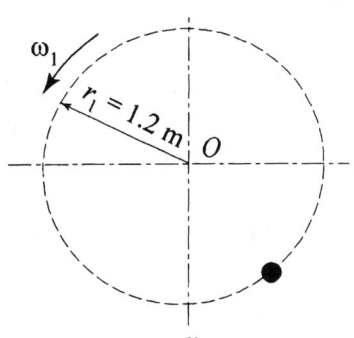
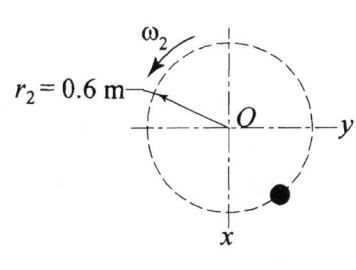

(a) FBD
(arbitrary time)

(b) Momentum Diagram
(time t_1)

(c) Momentum Diagram
(time t_2)

(2) Draw the momentum diagram at time t_1 in Fig. (b) above. Calculate the initial angular momentum about the z-axis: $(h_z)_1 = $ _____

(3) Draw the momentum diagram at time t_2 in Fig. (c) above. Calculate the final angular momentum about the z-axis: $(h_z)_2 = $ _____

(4) In the box below, use the angular impulse-momentum principle to find ω_2.

$(A_z)_{1\text{-}2} = (h_z)_2 - (h_z)_1$

$\omega_2 = $ _____ **Ans.**

Lesson 13 Kinematics of Relative Motion

Text Reference: Articles 15.1 and 15.2; Sample Problem 15.1

A.1 SELF-TEST (*To be done after assigned reading has been completed.*)

1. Motion that is described relative to a fixed reference frame is referred to as (a) _____ motion. Motion that is described relative to a moving reference frame is referred to as (b) _____ motion.

2. Complete the vector equation that relates each group of the terms listed in parts (a)-(d).

    (a) $\mathbf{r}_{B/A}, \mathbf{r}_{A/B}$: $\mathbf{r}_{B/A} =$

    (b) $\mathbf{v}_A, \mathbf{v}_B, \mathbf{v}_{B/A}$: $\mathbf{v}_B =$

    (c) $\mathbf{a}_{B/A}, \mathbf{a}_{A/B}$: $\mathbf{a}_{A/B} =$

    (d) $\mathbf{a}_A, \mathbf{a}_B, \mathbf{a}_{B/A}$: $\mathbf{a}_B =$

3. The base vectors for a translating reference frame can have non-zero derivatives. (T or F) ____

4. The base vectors for a rotating reference frame can have non-zero derivatives. (T or F) ____

A.2 General Comment

Remember that in this chapter only the special case of *translating* reference frames is considered.

B. GUIDED PROBLEM

PROBLEM 13.1

The two cars A and B are moving along straight, level roads with the constant speeds $v_A = 45$ mi/h and $v_B = 35$ mi/h, both directed as shown. When time $t = 0$, A is passing through

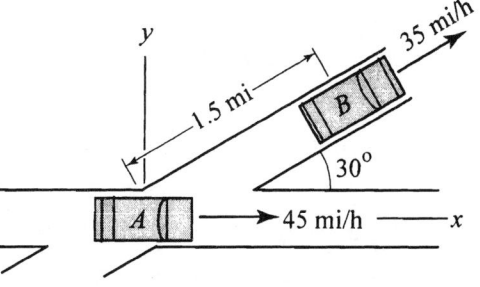

the intersection and B is 1.5 miles from the intersection. Determine $\mathbf{v}_{B/A}$ (velocity of B relative to A), and $\mathbf{r}_{B/A}$ (position vector of B relative to A) as functions of time.

a. Comments and Analysis

- The steps that we will use in the analysis are:

    (1) Using the xy-coordinate system shown, write the velocity vectors $\mathbf{v}_A$ and $\mathbf{v}_B$, and then compute $\mathbf{v}_{B/A} = \mathbf{v}_B - \mathbf{v}_A$. (Note that $\mathbf{v}_{B/A}$ will be time-independent because the velocity of each car is constant.)

    (2) Determine $\mathbf{r}_{B/A}$ by performing the time integral of $\mathbf{v}_{B/A}$. Evaluate the constant of integration by applying the initial condition at $t = 0$.

b. Guided Solution

(1) Write the vectors $\mathbf{v}_A$ and $\mathbf{v}_B$ using the xy-axes in the above problem figure:

$\mathbf{v}_A =$ _____

$\mathbf{v}_B =$ _____

Compute $\mathbf{v}_{B/A} = \mathbf{v}_B - \mathbf{v}_A =$ _____

   With the final result being $\mathbf{v}_{B/A} =$ _____ **Ans.**

(2) Integrate the relative velocity, letting $\mathbf{r}_0$ be the constant of integration:

$\mathbf{r}_{B/A} = \int \mathbf{v}_{B/A}\, dt =$ _____

State the initial condition:

   At $t = 0$, $\mathbf{r}_{B/A} = \mathbf{r}_0 =$ _____

Write the final expression for the relative position vector:

   $\mathbf{r}_{B/A} =$ _____ **Ans.**

Lesson 14 Kinematics of Constrained Motion

Text Reference: Article 15.3; Sample Problems 15.2 and 15.3

A. SELF-TEST (*To be done after assigned reading has been completed.*)

1. Geometric restrictions imposed on the motion of particles are referred to as _____.

2. *Equations of constraint* are mathematical expressions that describe the (a) _____ constraints on particles in terms of their (b) _____ coordinates.

3. Position coordinates of particles that are not subjected to kinematic constraints are called _____ coordinates.

4. Five coordinates are used to describe the configuration of a given system of particles. However, only three of the coordinates are kinematically independent. How many degrees of freedom does the system possess? ____

5. For a given (holonomic) mechanical system, let A = number of degrees of freedom, B = number of kinematic constraints, and C = number of position coordinates. Complete the equation that relates A, B and C: $A =$ _____

B. GUIDED PROBLEM

PROBLEM 14.1

Collar A is sliding downward along the vertical rod with the speed $v_A = 0.8$ m/s. The collar is connected to block B by an inextensible cable that passes over the pulley C. Find v_B, the velocity of block B, when $y_A = 1.2$ m.

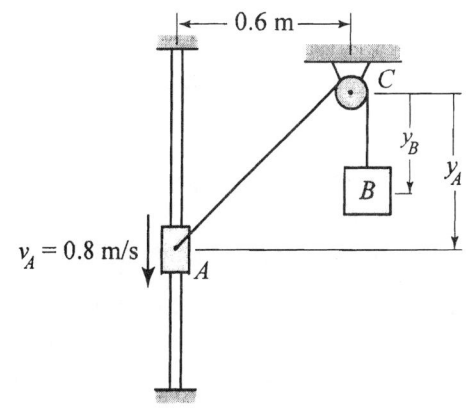

a. Comments and Analysis

- As shown in the figure, we will use the two position coordinates y_A and y_B to describe the configuration of the system.
- The constraint on the motion is that the cable does not change length.
- The steps that we will use in the analysis are:

    (1) Write the equation of constraint on the motion of the system.

    (2) Differentiate the equation of constraint to find v_B in terms of v_A and y_A.

    (3) Substitute the given conditions to evaluate v_B at the specified position.

b. Guided Solution

(1) Letting L be the length of the cable, write the equation of constraint:

$L = $ _____

(2) In the box below, differentiate the equation of constraint to find the relationship between v_B, v_A and y_A.

$$\frac{dL}{dt} = 0 =$$

which gives: $v_B = $ _____

(3) Substitute the given values to evaluate v_B at the specified position:

$v_B = $ _____ **Ans.**

Note: If desired, the relationship between the accelerations could be found by differentiating the expression for v_B with respect to time.

Lesson 15 Kinetics: Force-Mass-Acceleration Method

Text Reference: Article 15.4; Sample Problems 15.4–15.7

A.1 SELF-TEST *(To be done after assigned reading has been completed.)*

1. Why are constraint forces between particles eliminated when forces are summed over a closed system of particles?

   _____

2. Define each of the terms in the equation of motion for a *single* particle, $\Sigma \mathbf{F} = m\mathbf{a}$:

   $\Sigma \mathbf{F}$: _____

   m: _____

   $\mathbf{a}$: _____

3. Define each of the terms in the equation of motion for a closed system of particles, $\Sigma \mathbf{F} = m\bar{\mathbf{a}}$:

   $\Sigma \mathbf{F}$: _____

   m: _____

   $\bar{\mathbf{a}}$: _____

A.2 General Comment

The equation of motion of the mass center has limited application because it gives no information about the motion of the individual particles. Therefore, the equations of motion for the individual particle are often analyzed.

B. GUIDED PROBLEM

PROBLEM 15.1

The 20-kg block B rests on the 50-kg block A. The blocks are being pushed up the frictionless inclined plane by the force $P = 500$ N that is parallel to the plane. Determine the smallest coefficient of static friction μ for the surface between the blocks that will prevent B from sliding on A.

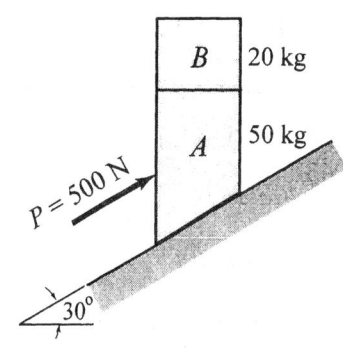

a. Comments and Analysis

- The accelerations of blocks are the same; there is no relative motion between them.
- The smallest coefficient of friction that will prevent sliding occurs when motion impends on the surface between the blocks.
- The steps that we will use in the analysis are:
    (1) Calculate the acceleration of the blocks by analyzing the system of both blocks.
    (2) Determine the smallest coefficient of friction to prevent sliding by analyzing block B when motion impends between the blocks.

b. Guided Solution

(1) On the figures below, draw the FBD and MAD for the system of both blocks. Label the normal force as N_A and the acceleration as a. Then, calculate the acceleration by writing and solving the FMA equation in the x'-direction.

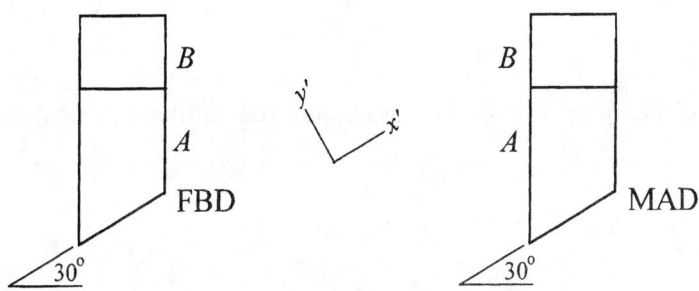

$\Sigma F_{x'} = ma_{x'}$ _____ gives $a =$ _____

(continued)

(2) On the figures below, draw the FBD and MAD for block B. Label the normal force as N_B, the friction force as F_B, and use the value for the acceleration as a that was found in Step (1).

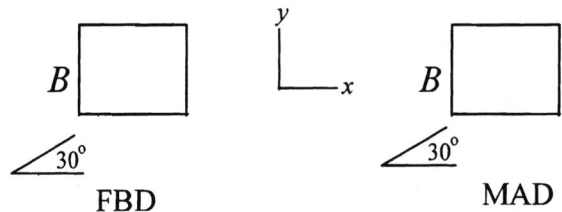

FBD   MAD

Write the FMA equation in the y-direction and solve for N_B:

$\Sigma F_y = ma_y$: $+\uparrow$ _____ gives $N_B =$ _____

Write the FMA equation in the x-direction and solve for F_B:

$\Sigma F_x = ma_x$ $\xrightarrow{+}$ _____ gives $F_B =$ _____

Using $F_B = \mu N_B$ (motion impends), solve for μ:

$\mu =$ $\mu =$ _____ ANS.

Lesson 16 Work-Energy and Impulse-Momentum Principles

Text Reference: Articles 15.5–15.7; Sample Problems 15.8–15.11

Note: *This Lesson extends the work-energy and impulse-momentum (linear and angular) principles for a single particle to systems of particles.*

A. SELF-TEST (*To be done after assigned reading has been completed.*)

1. Identify each term in the work-energy principle for a closed system of particles where 1 and 2 refer to the initial and final positions of the system, respectively:

    $(U_{1-2})_{ext} + (U_{1-2})_{int} = \Delta T$.

    (a) $(U_{1-2})_{ext}$: _____

    (b) $(U_{1-2})_{int}$: _____

    (c) ΔT: _____

2. List two types of internal forces that can do work on a system of particles:

    _____ and _____

3. For the mechanical energy of a system to be conserved, both internal and external forces must be conservative. (T or F) _____

4. Identify each term in the linear momentum equation for a system of particles: $\mathbf{p} = m\bar{\mathbf{v}}$.

    (a) $\mathbf{p}$: _____

    (b) m: _____

    (c) $\bar{\mathbf{v}}$: _____

5. In the linear impulse-momentum equation for a system of particles, $\mathbf{L}_{1-2} = \Delta \mathbf{p}$, $\mathbf{L}_{1-2}$ represents the linear impulse of both external and internal forces. (T or F) _____

(continued)

6. If the momentum of a system of particles is conserved, the momentum of each particle of the system is also conserved. (T or F) _____

7. List the two choices for point A for which the moment-angular momentum equation, $\Sigma \mathbf{M}_A = \dot{\mathbf{h}}_A$ is valid. _____ and _____

8. Consider the angular impulse-momentum principle for a system of particles:
$$(\mathbf{A}_A)_{1-2} = (\mathbf{h}_A)_2 - (\mathbf{h}_A)_1$$
   (a) The term $(\mathbf{A}_A)_{1-2}$ refers to the angular impulse of external forces. (T or F) _____
   (b) The equation is valid for any choice of point A. (T or F) _____

B. GUIDED PROBLEMS

PROBLEM 16.1

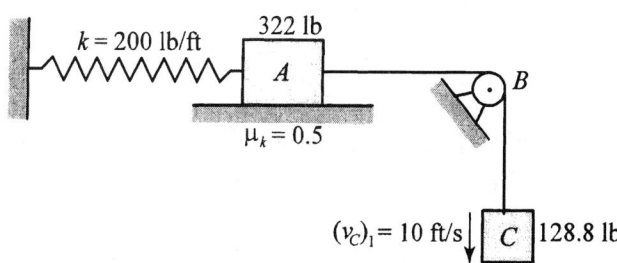

Block A weighing 322 lb is connected by an inextensible cable ($k = 200$ lb/ft) to block C that weighs 128.8 lb. The weight of pulley B can be neglected. The coefficient of kinetic friction between A and the plane is $\mu_k = 0.5$. In the position shown, C is moving downward with the velocity $(v_C)_1 = 10$ ft/s and the spring is stretched 2 ft. Determine $(v_C)_2$, the velocity of C after it has moved down 0.8 ft from the position shown.

a. Comments and Analysis

- We will use the work-energy principle to solve this problem. We identify position 1 to be the initial position shown, and position 2 to be the final position.

- We consider our system to consist of the two blocks connected by the cable and the pulley. The FBD showing all the forces that act on the system is shown below. Careful study of this FBD reveals that external work will be done on the system by F_{spring}, W_C, and F_A (the friction force). The following external forces do not do work on the system: W_A and N_A (because both forces are perpendicular to the displacement of A), and the pin reaction R acting on the pulley (this force does not move). The cable tension, an internal force, does not do work because the cable is inextensible.

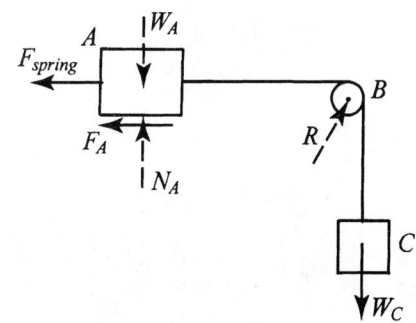

- The steps that we will use in the analysis are:

    (1) Compute the total work done on the system.

    (2) Evaluate T_1, the initial kinetic energy.

    (3) Write the expression for T_2, the final kinetic energy.

    (4) Using the above results, apply the work-energy principle to compute $(v_C)_2$.

(continued)

b. Guided Solution

(1) Compute the total work done on the system.

    (a) Work done by weight of block C.

        $(U_{1\text{-}2})_{ext} = Wh =$ _____

    (b) Work done by the spring.

        The spring deformations are: $\delta_1 =$ _____ and $\delta_2 =$ _____

        $(U_{1\text{-}2})_{ext} = -\dfrac{1}{2}k(\delta_2^2 - \delta_1^2) =$ _____

    (c) Work done by the friction force.

        Draw the FBD of block A on the figure at the right. Calculate N_A, and then compute F_A.

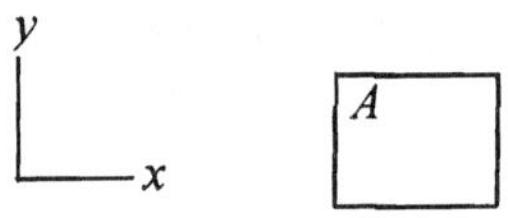

        $\Sigma F_y = ma_y: \;+\uparrow$ _____ gives $N_A =$ _____

        Therefore, the friction force $F_A =$ _____

        $(U_{1\text{-}2})_{ext} = -F_A d =$ _____

(2) Evaluate T_1, the initial kinetic energy.

$$T_1 = \dfrac{1}{2}m_A(v_A)_1^2 + \dfrac{1}{2}m_C(v_C)_1^2$$

$T_1 =$ _____

(3) Write the expression for T_2, the final kinetic energy [use $(v_A)_2 = (v_C)_2$]

$$T_2 = \dfrac{1}{2}m_A(v_A)_2^2 + \dfrac{1}{2}m_C(v_C)_2^2$$

$T_2 =$ _____

(4) Using the above results, in the space below apply the work-energy principle and compute $(v_C)_2$.

        $(U_{1\text{-}2})_{ext} + (U_{1\text{-}2})_{int} = T_2 - T_1$

$(v_C)_2 =$ _____ **Ans.**

PROBLEM 16.2

The 40-lb block A and the 24-lb block B are attached to the ends of a linear spring with constant $k = 50$ lb/in. and unstretched

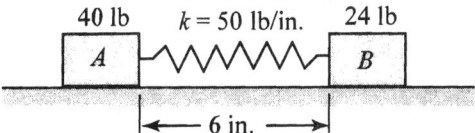

length $L_0 = 10$ in. The blocks are released from rest in the position shown where the length of the spring is 6 in. Neglecting friction, compute the velocity of each block when the length of the spring is 11 in.

a. Comments and Analysis

Our solution of this problem will use both the impulse-momentum principle and the work-energy principle. We will let subscripts 1 and 2 refer to the initial and final positions and corresponding initial and final times.

- The steps that we will use in the analysis are:

    (1) Draw the free-body diagram (FBD) of the system in an arbitrary position.

    (2) Draw the momentum diagram of the entire system at the final time t_2.
       (There is no momentum at the intial time t_1.)

    (3) Identify all of the unknowns on the FBD and momentum diagram.

    (4) Write the impulse-momentum equation for the system.

    (5) Write the work-energy equation for the system (only the spring does work).

    (6) Solve the equations for the final velocity of each block.

b. Guided Solution

(1) On the figure below, draw the FBD of the entire system in an arbitrary position.
    Label the weights as W_A and W_B, and the normal forces as N_A and N_B.

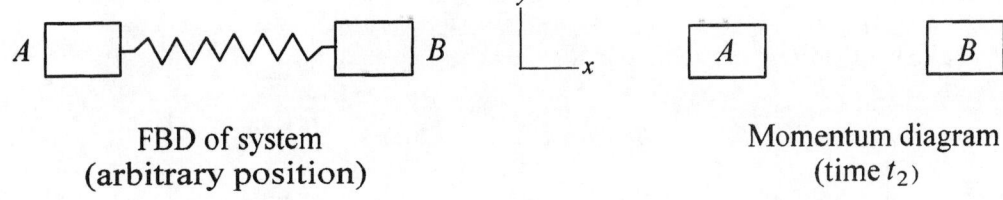

FBD of system
(arbitrary position)

Momentum diagram
(time t_2)

(2) On the figure above, draw the momentum diagram at the final time t_2.

Label the masses as m_A and m_B, and assume the final velocity $(v_A)_2$ is to the left and the final velocity $(v_B)_2$ is to the right.

(continued)

(3) Identify all of the unknowns that appear on the FBD and momentum diagram.

    (a) From the FBD of each block (not shown here), the equation $\sum F_y = 0$ would determine that $N_A = W_A$ and $N_B = W_B$. Therefore, the following are the only two unknowns remaining on the two diagrams: _____ and _____ .

(4) In the space below, refer to the above diagrams and write the impulse-momentum equation in the x-direction for the system. Substitute the numerical to obtain the relationship between the final velocities.

$$(L_{1\text{-}2}) = (p_x)_2 - (p_x)_1$$

$\xrightarrow{+}$

                      The relationship between the final velocities: $(v_A)_2 =$ ____$(v_B)_2$ **Eq. (1)**

(5) Write the work-energy equation for the system.

    (a) Work done by the spring.

        Deformations: $\delta_1 = L_1 - L_0 =$ _____ $\delta_2 = L_2 - L_0 =$ _____

$$(U_{1\text{-}2}) = -\frac{1}{2}k(\delta_2^2 - \delta_1^2) = \underline{\hspace{2cm}} = \underline{\hspace{1.5cm}}\text{ lb·ft} = \underline{\hspace{1.5cm}}\text{ lb·in.}$$

    (b) Evaluate T_1, the initial kinetic energy: $T_1 =$ _____

    (c) Write the expression for T_2, the final kinetic energy:

        $T_2 =$ _____

    (d) In the space below, write the work-energy principle for the system. Substitute the numerical values to obtain the equation relating $(v_A)_2$ and $(v_B)_2$.

$$(U_{1\text{-}2}) = T_2 - T_1$$

Equation relating $(v_A)_2$ and $(v_B)_2$: _____ **Eq. (2)**

(continued)

(6) In the space below, solve Eqs. (1) and (2) for the velocity of each block.

$(v_A)_2 =$ _____ and $(v_B)_2 =$ _____ **Ans.**

PROBLEM 16.3

The figure shows two small balls A ($m_A = 2$ kg) and B ($m_B = 3$ kg) that are attached to the ends of the rigid rod of negligible mass. The assembly is initially at rest when the constant 100-N horizontal force, perpendicular to the rod, is applied at A. The force remains perpendicular to the rod as it rotates about the vertical z-axis at O. Determine the time required for the rod to reach an angular velocity of 90 rad/s. Neglect friction.

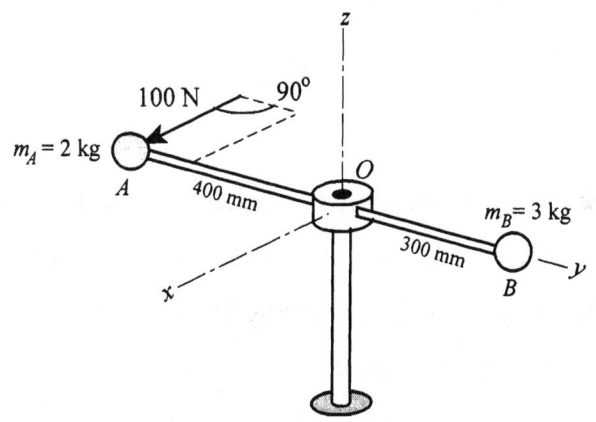

a. Comments and Analysis

- We will let $t_1 = 0$ refer to the time when the rotation starts and t_2 be the time when the angular velocity ω is 90 rad/s.
- We will solve this problem using the angular impulse-angular momentum method.
- The steps that we will use in the analysis are:

    (1) Draw the FBD of the assembly (valid throughout the time period).

    (2) Draw the momentum diagram of the assembly at the final time t_2. (There is no momentum at the initial time t_1.)

    (3) Write the angular impulse-angular momentum equation for the z-axis and solve for the time t_2 when the final angular velocity is reached.

(continued)

b. Guided Solution

(1) Draw the FBD of the assembly on the figure below (the 100-N force is already shown). Although the FBD is drawn when the assembly is in the initial position, it is stated that the 100-N force remains perpendicular to the rod throughout the motion.

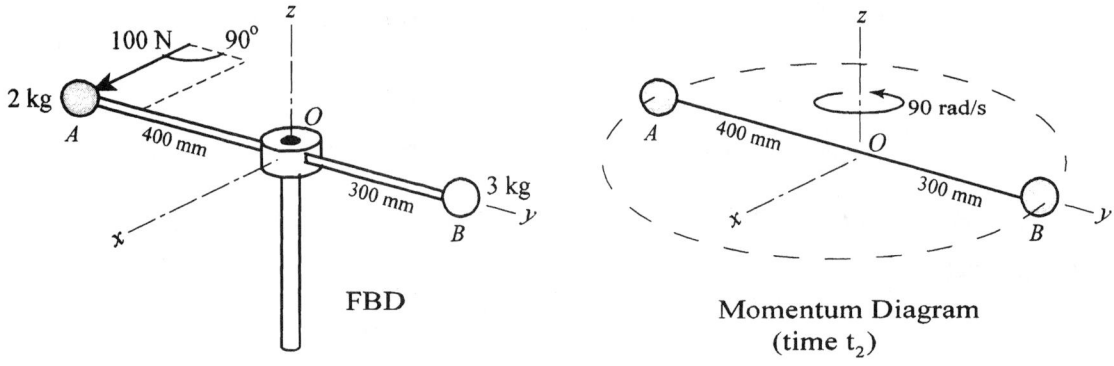

FBD

Momentum Diagram (time t_2)

(2) Draw the momentum diagram of the assembly at the final time t_2 on the figure above. Evaluate each term using the given numerical values.

(3) In the space below, write the angular impulse-angular momentum equation for the z-axis and solve for the time t_2 when the final angular velocity is reached. (Note that there is no momentum at the initial time t_1.)

$(A_z)_{1\text{-}2} = (h_z)_2 - (h_z)_1$

$t_2 = $ _____ **Ans.**

Lesson 17 Plastic Impact and Impulsive Motion

Text Reference: Articles 15.8 and 15.9; Sample Problems 15.12 and 15.13

A. SELF-TEST (*To be done after assigned reading has been completed.*)

1. The contact forces caused by the collision of two particles are called

   (a) _____ forces. What notation is used to indicate these forces on free-body diagrams? (b) _____

2. What is meant by stating that the impact of two particles is *plastic*?

   _____

3. Motion for which the time of impact $\Delta t \to 0$ is called _____ motion.

4. When the time of impact is infinitesimal for two impacting blocks, the following conclusions can be made for the impact interval:

   (a) the magnitudes of impact forces are _____

   (b) the impulses of finite forces are _____

   (c) the accelerations of the blocks are _____

   (d) positions of the blocks _____

5. The analysis of the impact of particles involves writing and solving the impulse-momentum equations using the following three types of diagrams:

   (a) _____ diagrams before impact

   (b) _____ diagrams during impact

   (c) _____ diagrams after impact

B. GUIDED PROBLEMS

PROBLEM 17.1

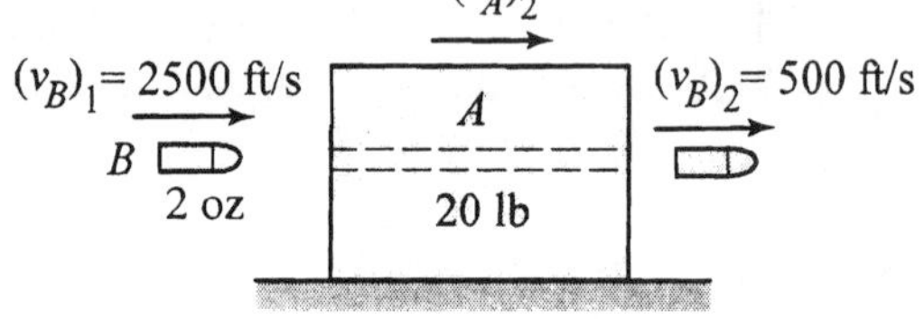

The 2-oz bullet B is fired at $(v_B)_1 = 2500$ ft/s into the stationary 20-lb block A. The bullet passes through the block and emerges with the horizontal velocity $(v_B)_2 = 500$ ft/s. Determine the velocity $(v_A)_2$ with which the block begins to move after the impact. Neglect friction.

a. Comments and Analysis

- Note that the subscripts 1 and 2 are being used to indicate the velocities immediately before and immediately after the impact, respectively.

- The steps that we will use in the analysis are:

    (1) Draw the FBDs of A and B during the impact.

    (2) Draw the momentum diagrams for A and B immediately before, and immediately after the impact.

    (3) Write the impulse-momentum equation for the system of A and B. Solve for the velocity $(v_A)_2$ with which the block begins to move after the impact.

b. Guided Solution

(1) On Fig. 1 below, draw the FBDs of A and B during the impact. Use a caret above the letter to indicate an impact force.

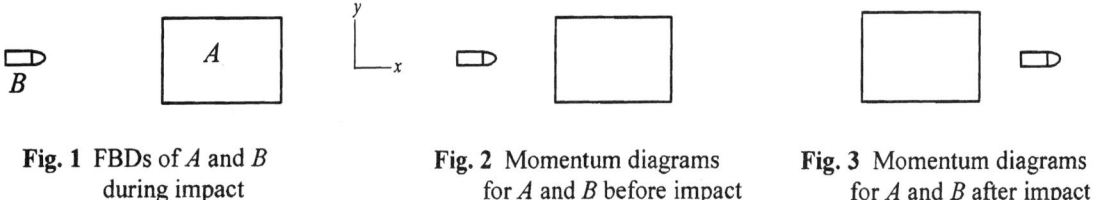

Fig. 1 FBDs of A and B during impact

Fig. 2 Momentum diagrams for A and B before impact (Use symbols not numbers)

Fig. 3 Momentum diagrams for A and B after impact (Use symbols not numbers)

(2) On Figs. 2 and 3 above, draw the momentum diagrams for A and B immediately before, and immediately after impact. For convenience, use symbols (m_A, m_B, $(v_A)_1$, $(v_B)_1$, and so on) not numbers.

(continued)

(3) In the space below, complete the solution by writing the impulse-momentum equation *for the system* and solving for $(v_A)_2$.

$(L_{1\text{-}2}) = (p_x)_2 - (p_x)_1$

$\xrightarrow{+}$

$(v_A)_2 = $ _____ **Ans.**

c. Concluding Remarks

1. The analysis has shown that there are no forces acting *on the system* in the x-direction. Therefore, it was not necessary to assume that the time of impact was negligible.
2. The impulse of the impact force that acted between A and B could be computed by writing the impulse-momentum equation for either A or B.

PROBLEM 17.2

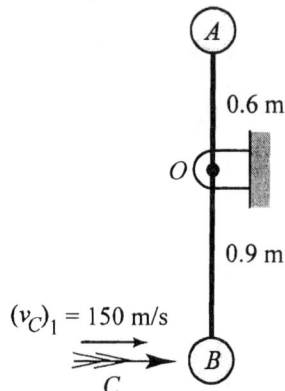

The assembly consists of two small balls A and B that are connected to a light slender rod which is pinned at O. The assembly is initially at rest when the dart C is fired into B with the velocity $(v_C)_1 = 150$ m/s. Determine the angular velocity ω_2 with which the bar begins to rotate. Assume that the assembly lies in the vertical plane. Neglect the duration of the impact. The masses are: $m_A = 0.75$ kg, $m_B = 1.50$ kg, and $m_C = 0.075$ kg.

a. Comments and Analysis

- Note that the subscripts 1 and 2 are being used to indicate the velocities immediately before, and immediately after the impact, respectively.

- The steps that we will use in the analysis are:

    (1) Draw the FBDs of C and the bar during the impact.

    (2) Draw the momentum diagrams for C and the bar immediately before, and immediately after, the impact.

    (3) Write the impulse-momentum equation for the assembly. Solve for the angular velocity ω_2 with which the bar begins to rotate

b. Guided Solution

(1) On Fig. 1 below, draw the FBDs of C and the bar during the impact. Label the unknowns as forces P, O_x and O_y (with a caret above each letter to indicate that it could be an impulsive force).

(continued)

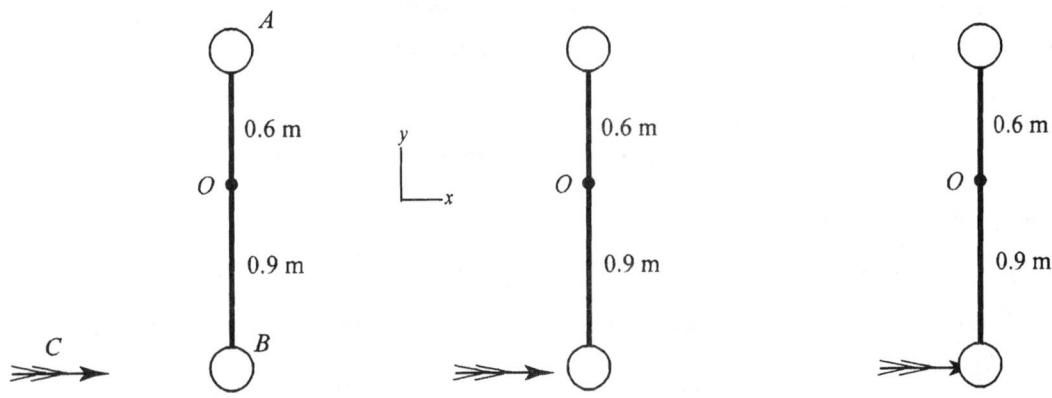

Fig. 1 FBDs during impact   **Fig. 2** Momentum diagrams before impact   **Fig. 3** Momentum diagrams after impact

(2) On Figs. 2 and 3 above, draw the momentum diagrams for C and the bar immediately before, and immediately after, the impact.

Note: Referring to Figs. 1-3, you will see that there are a total of four unknowns: the impulses of the three forces $\hat{P}, \hat{O}_x,$ and $\hat{O}_y$, and the final angular velocity ω_2.

There are also four impulse-momentum equations: three for the bar and one for the dart. Therefore, the four equations could be solved for the four unknowns. However, the angular velocity ω_2 can be found using only one equation.

(3) In the space below, complete the solution by writing the angular impulse-angular momentum equation about point O *for the assembly* and solving for ω_2.

$(A_O)_{1\text{-}2} = (h_O)_2 - (h_O)_1$

↻(+)

$\omega_2 =$ _____ **Ans.**

c. Concluding Remark

Note that the answer for ω_2 would be the same if the assembly were in the horizontal plane instead of the vertical plane, as stated.

Lesson 18 Elastic Impact

Text Reference: Article 15.10; Sample Problems 15.14–15.16

A. SELF-TEST (*To be done after assigned reading has been completed.*)

1. The property of a body to return either totally or partially to its original shape is called _____ .

2. When two particles collide, the line that is perpendicular to the contact surface is called the _____ .

3. When two particles collide with both velocities directed along the line of impact, the impact is called (a) _____ impact; otherwise the impact is known as (b) _____ impact.

4. The experimental constant *e* that characterizes the "elasticity" of colliding bodies is called the _____ .

5. What is the value of *e* when no energy is lost during the impact? _____

6. What is the value of *e* when the impact is plastic? _____

7. Two particles *A* and *B* collide with direct impact. Letting v_{sep} be their velocity of separation and v_{app} be their velocity of approach, write the defining equation for *e*:

    $e =$

8. Two particles *A* and *B* collide with oblique impact. Which components of their velocities are used in the defining equation for *e*?

    _____

B. GUIDED PROBLEM

PROBLEM 18.1

The 5-lb disk A and the 10-lb disk B are sliding across the horizontal plane with the initial velocities $(v_A)_1 = 30$ ft/s and $(v_B)_1 = 10$ ft/s, directed as shown. The coefficient of restitution for the impact is $e = 0.5$. Find the velocities $(v_A)_2$ and $(v_B)_2$ immediately after the impact. Neglect friction.

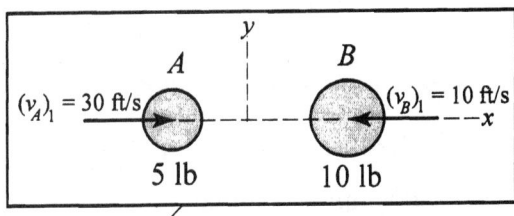

a. Comments and Analysis

- Note that the subscripts 1 and 2 are being used to indicate the velocities immediately before and immediately after the impact, respectively.
- The steps that we will use in the analysis are:

    (1) Draw the FBDs of A and B during the impact showing only forces in the horizontal plane.

    (2) Draw the momentum diagrams for A and B immediately before, and immediately after the impact.

    (3) Write the impulse-momentum equation for the system of A and B.

    (4) Write the coefficient of restitution equation relating the initial and final velocities.

    (5) Solve the two equations for the final velocities $(v_A)_2$ and $(v_B)_2$.

b. Guided Solution

(1) On Fig. 1 below, draw the FBDs of A and B during the impact showing only forces in the horizontal plane. Use a caret above the letter to indicate an impact force.

 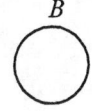   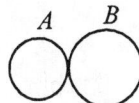   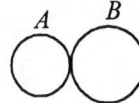

Fig. 1 FBD's during impact (forces acting in *xy*-plane only)   **Fig. 2** Momentum diagram before impact   **Fig. 3** Momentum diagram after impact

(continued)

(2) On Figs. 2 and 3 above, draw the momentum diagrams for A and B immediately before, and immediately after impact. For convenience, use symbols (m_A, m_B, $(v_A)_1$, $(v_B)_1$, and so on) not numbers.

(3) Referring to Figs. 2 and 3, in the space below, write the impulse-momentum equation in the x-direction for the system of A and B. Substitute the known numerical values and simplify your equation.

$(L_{1\text{-}2}) = (p_x)_2 - (p_x)_1$
$\xrightarrow{+}$

Impulse-momentum equation is: _____ **Eq. (1)**

(4) In the space below, write and simplify the coefficient of restitution equation.

$$e = \frac{v_{sep}}{v_{app}}$$

Coefficient of restitution equation is: _____ **Eq. (2)**

(5) In the space below, solve Eqs. (1) and (2) for the final velocities.

$(v_A)_2 =$ _____ and $(v_B)_2 =$ _____ **Ans.**

Lesson 19 Plane Angular Motion; Rotation about a Fixed Axis

Text Reference: Article 16.1–16.3; Sample Problems 16.1–16.3

A. SELF-TEST (*To be done after assigned reading has been completed.*)

1. Why is the rigid body concept an "idealization"?

   _____

2. Motion for which all points in a body remain a constant distance from a fixed reference plane is called _____ motion

3. List the three categories of plane motion.
   (a) _____ ; (b) _____ ; (c) _____

4. A rigid body is in plane motion with *AB* being a line in the body and in the plane of motion. The angle $\theta(t)$ between *AB* and a fixed reference line (in the plane of motion) is called the _____ of line *AB*.

5. Refer to the body described above in Question 4. If $\theta(t)$ changes to $\theta(t + \Delta t)$ in the time interval Δt, define the angular displacement of *AB* during the time interval:

   $\Delta \theta =$

6. Refer to the body described above in Question 4. Why is it meaningful to refer to the angular displacement of line *AB* as the angular displacement *of the body*?

   _____

7. For a body undergoing plane motion, the time derivative of the angular position coordinate $\theta(t)$ is defined to be the (a) _____ of the body, and it is indicated by the Greek letter (b) ____.

(continued)

8. For a body undergoing plane motion, the time derivative of the angular velocity of the body is defined to be the (a) _____ of the body, and it is indicated by the Greek letter (b) ____ .

9. When a rigid body is undergoing three-dimensional motion:
   (a) the angular displacement of the body is a vector (T or F) ____
   (b) the angular velocity of the body is a vector (T or F) ____
   (c) the angular acceleration of the body is a vector (T or F) ____

10. A body is rotating about a fixed axis. Describe the path followed by an arbitrary point on the body that is located a distance R from the axis of rotation:
    _____

11. Complete the following *scalar* equations for the velocity (v) and the acceleration components (a_n, a_t) of a point in a rigid body that is rotating about a fixed axis. The distance of the point to the axis is R; and the angular velocity and angular acceleration of the radial line to the point are ω and α, respectively.

    (a) $v = R(\quad)$; (b) $a_n = R(\quad) = \dfrac{(\quad)}{R} = v(\quad)$; (c) $a_t = R(\quad)$

12. Complete the following *vector* equations for the velocity and acceleration components of a point in a rigid body that is rotating about a fixed axis. The position vector of the point relative to the fixed axis is **r** and the velocity of the point is **v**. Furthermore, the angular velocity vector and angular acceleration vector of the radial line to the point are **ω** and **α**, respectively.

    (a) $\mathbf{v} = (\quad) \times \mathbf{r}$; (b) $\mathbf{a}_n = \boldsymbol{\omega} \times (\quad) = (\quad) \times (\quad \times \mathbf{r})$; (c) $\mathbf{a}_t = (\quad) \times \mathbf{r}$

B. GUIDED PROBLEMS

PROBLEM 19.1

The angular acceleration α of a line rotating in a plane is $\alpha = 12t^2 + 2k$ rad/s^2, where t is the time in seconds, k (rad/s^2) is a constant, and CCW angles are positive. When $t = 0$, the angular position θ of the line from a fixed reference line is 2 radians CW and the angular velocity ω of the line is 3 rad/s CCW. When $t = 1.0$ s, the angular position of the line is 4 rad CW. Determine the angular acceleration of the line when $t = 2.0$ s.

a. Comments and Analysis

- The steps that we will use in the analysis are:

    (1) Write each of the stated conditions in equation form.

    (2) Integrate the angular acceleration α to determine the angular velocity ω, then integrate ω to obtain the angular position coordinate θ.

    (3) Apply the stated conditions to evaluate the constants of integration and the constant k. Having determined k, the angular acceleration will be known.

b. Guided Solution

(1) Write the stated conditions in terms of θ and ω.

Referring to the problem statement, the stated conditions are (be sure to use the proper signs):

1. at $t = 0$, $\theta =$ _____ **2.** at $t = 0$, $\omega =$ _____ **3.** at $t = 1.0$ s, $\theta =$ _____

(2) Integrate $\alpha = 12t^2 + 2k$ with respect to time once to obtain ω, and a second time to obtain θ. Let the constants of integration be C_1 and C_2.

$$\omega = \int \alpha \, dt =$$

$$\theta = \int \omega \, dt =$$

(3) Note that there are 3 stated conditions and 3 unknown constants. In the space below, evaluate the 3 constants, and then evaluate α when $t = 2$ s.

$\alpha_{(t=2\text{ s})} =$ _____ **Ans.**

PROBLEM 19.2

The friction disks A and B rotate in contact with each other without slipping. At a certain instant of time, B is rotating with the angular velocity $\omega_B = 2$ rad/s and angular acceleration $\alpha_B = 6$ rad/s², both clockwise.

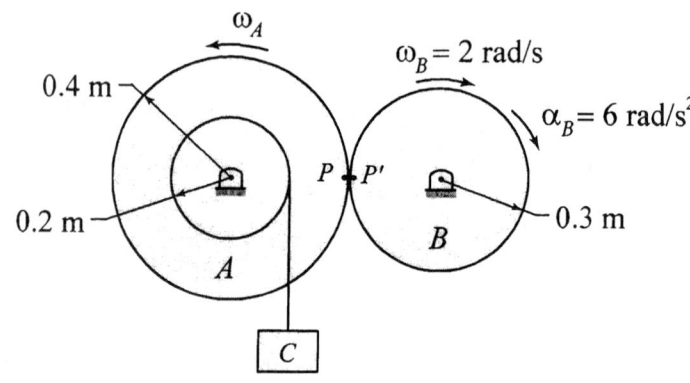

At the same instant, A is rotating counterclockwise with the angular velocity ω_B. The cable wrapped around the hub of A causes block C to move upward. Determine v_C and a_C, the velocity and acceleration of block C.

a. Coments and Analysis

- We identify the two points that are in contact: P on disk A and P' on disk B. Note that the paths of these points are circles.

b. Guided Solution

(1) Calculate the velocity of point P': $v_{P'} = $ _____

(2) Because there is no slipping, what can we conclude about the velocities of P and P'?
_____

(3) Using the result of (2) above, the velocity of point P is $v_P = $ _____

(4) Calculate $\omega_A = $ _____

(5) Calculate $v_C = $ _____ **Ans.**

(6) Because there is no slipping, what can we conclude about the accelerations of P and P'? _____

(7) Calculate $(a_{P'})_t = $ _____

(8) Using the result of (7) above, for point P we have $(a_P)_t = $ _____

(9) Calculate $\alpha_A = $ _____

(10) Calculate $a_C = $ _____ **Ans.**

Lesson 20 Relative Motion; Method of Relative Velocity

Text Reference: Articles 16.4 and 16.5; Sample Problems 16.4–16.6

A. SELF-TEST (*To be done after assigned reading has been completed.*)

1. When a rigid body is undergoing three-dimensional motion:
   (a) the angular displacement of the body is a vector (T or F) ____
   (b) the angular velocity of the body is a vector (T or F) ____
   (c) the angular acceleration of the body is a vector (T or F) ____

2. For general plane motion, the relative velocity and relative acceleration of point B relative to point A (A and B belong to the same rigid body) can be written using the equations for rotation about a fixed axis, with the following changes in notation:

   $r_{B/A}$ replaces R ; $v_{B/A}$ replaces v ; $(a_{B/A})_n$ replaces a_n; $(a_{B/A})_t$ replaces a_t

   Using the appropriate changes, write the equivalent relative motion equations for each of the equations below for rotation about a fixed axis:

Fixed axis rotation	Motion of point B relative to point A
Scalar equations	
Velocity: $v = R\omega$	(a) Relative velocity: $v_{B/A} =$
Normal component of acceleration: $a_n = R\omega^2$	(b) Normal component of relative acceleration: $(a_{B/A})_n =$
Tangential component of acceleration: $a_t = R\alpha$	(c) Tangential component of relative acceleration: $(a_{B/A})_t =$
Vector equations	
$\mathbf{v} = \boldsymbol{\omega} \times \mathbf{r}$	(d) $\mathbf{v}_{B/A} =$
$\mathbf{a}_n = \boldsymbol{\omega} \times (\boldsymbol{\omega} \times \mathbf{r})$	(e) $(\mathbf{a}_{B/A})_n =$
$\mathbf{a}_t = \boldsymbol{\alpha} \times \mathbf{r}$	(f) $(\mathbf{a}_{B/A})_t =$

(*continued*)

3. There can be relative motion between two points that belong to a rigid body that is translating. (T or F) _____

4. (a) If A and B are two points in the same rigid body that is undergoing plane motion, how many variables are in the relative velocity $\mathbf{v}_B = \mathbf{v}_A + \mathbf{v}_{B/A}$? _____

   (b) How many variables must be known before this relative velocity equation can be solved? _____

5. A point for which the magnitude and/or direction of its velocity is known is referred to as a _____.

6. If a point is undergoing plane motion along a given path, why does its velocity contain at most one unknown? _____.

7. List the steps in the application of the relative velocity method:

   Step 1: Identify two (a) _____ points for velocity.

   Step 2: Write the (b) _____ equation between the points.

   Step 3: Solve the equation if there are how many unknowns? (c) _____

8. A wheel of radius R rolls without slipping with angular velocity ω and angular acceleration α. Write the equations for the magnitudes of the velocity v_O and acceleration a_O of the center O of the wheel:

   (a) $v_O = $ _____ ; (b) $a_O = $ _____

   The direction of v_O must be consistent with the direction of (c) _____.

   The direction of a_O must be consistent with the direction of (d) _____.

B. GUIDED PROBLEMS

PROBLEM 20.1

- The spool rolls without slipping on the horizontal plane with slipping. In the position shown, the velocity of point A on the spool is 6 ft/s to the right. Determine the angular velocity of the spool for this position.

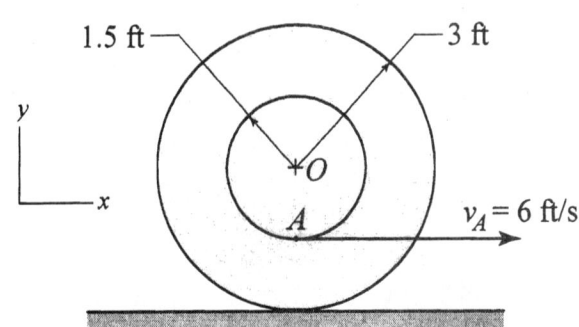

Solve using both scalar notation (Solution I) and vector notation (Solution II).

a. Comments and Analysis

- Because the spool rolls without slipping, the velocity of it center O is known to be horizontal of magnitude $r\omega$, with its sense consistent with the direction of ω.

- We will assume that the angular velocity ω of the spool is clockwise. (This is an assumption but you may see by inspection of the figure that clockwise will be the correct direction.)

- The steps that we will use in the analysis are:

    (1) Identify the kinematically important points.

    (2) Write the relative velocity equation between the kinematically important points, and solve for the equation for the angular velocity ω. (We will accomplish this step using both scalar and vector notation.)

b. Guided Solution

(1) Identify the kinematically important points.

   (i) A is a kinematically important point because _____

   (ii) O is a kinematically important point because _____

Writing the relative velocity equation between points A and O will be straightforward because both points belong to the same rigid body.

(continued)

Solution I *(using scalar notation)*

(2) Complete the following relative velocity equation between *A* and *O* by placing each of the terms in the dotted boxes using scalar notation.

$$\mathbf{v}_A = \mathbf{v}_O + \mathbf{v}_{A/O}$$

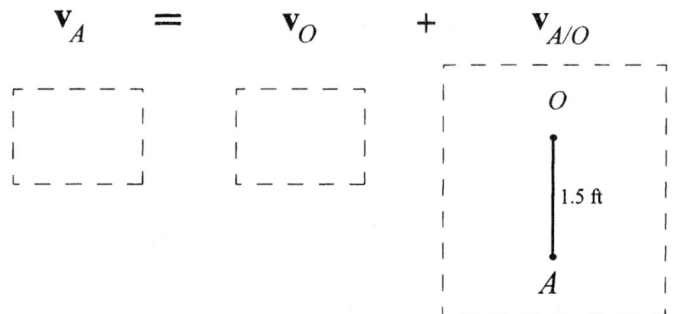

In the space below, solve the above equation for the angular velocity. (Express your answer using vector notation.)

ω = _____ Ans.

Solution II *(using vector notation)*

(2) Write the relative velocity equation between *A* and *O* using vector notation.

The relative velocity equation between points *A* and *O* is

$$\mathbf{v}_A = \mathbf{v}_O + \mathbf{v}_{A/O} = \mathbf{v}_O + \boldsymbol{\omega} \times \mathbf{r}_{A/O} \qquad (a)$$

Write each of the following terms using vector notation. ω

$\mathbf{v}_A =$ _____ $\mathbf{v}_O =$ _____

ω (unknown, assumed to be clockwise) = _____

$\mathbf{r}_{A/O} =$ _____ (refer to the problem figure)

$\boldsymbol{\omega} \times \mathbf{r}_{A/O} =$ _____

In the space below, substitute the above terms into Eq. (a) and solve for the angular velocity (Express your answer using vector notation.)

ω = _____ Ans.

PROBLEM 20.2

In the position shown, the angular velocity of bar AB is 4 rad/s clockwise. For this position, determine the angular velocities or bars BC and CD. Solve using scalar notation (Solution I) and vector notation (Solution II).

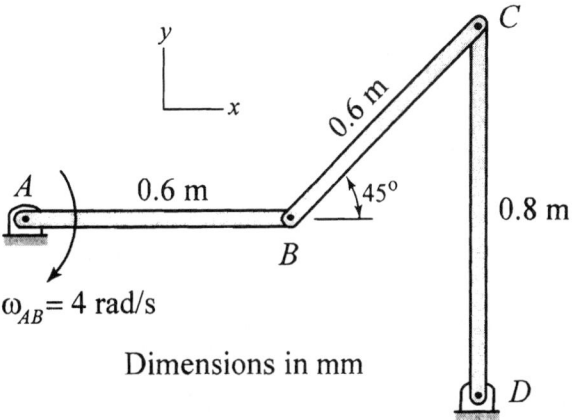

a. Comments and Analysis

- You would probably be able to determine the correct directions for the angular velocities of BC and CD by inspection of the problem figure. However, without attempting to do this, we will simply assume that both directions are counterclockwise, and rely upon the signs in our answers to tell us the correct directions.

- The steps that we will use in the analysis are:
    (1) Identify the kinematically important points.
    (2) Write the relative velocity equation between the appropriate kinematically important points, and solve for the equation for the angular velocities ω_{BC} and ω_{CD}. We will accomplish this step using both scalar and vector notation.)

b. Guided Solution

(1) Identify the kinematically important points.

   (i) A and D are kinematically important points because _____

   (ii) B is a kinematically important point.

   What is the path of B? _____

   (iii) C is a kinematically important point.

   What is the path of C? _____

Writing the relative velocity equation between points B and C will be straightforward because both points belong to the same rigid body.

(continued)

Solution I *(using scalar notation)*

(2) Complete the following relative velocity equation between B and C by placing each of the terms in the dotted boxes using scalar notation.

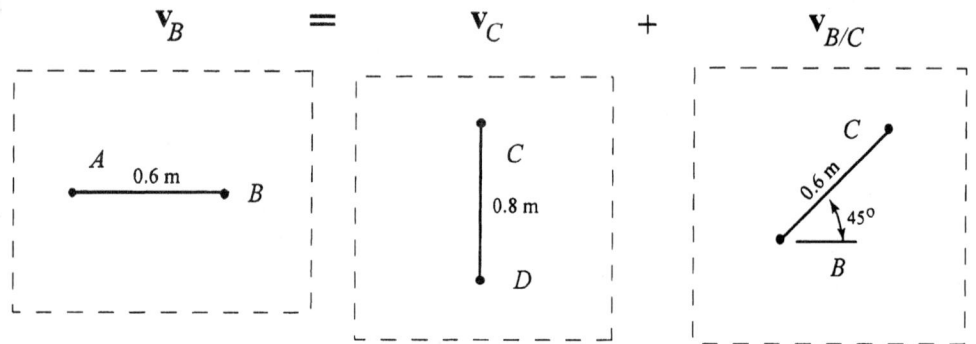

In the space below, solve the above equation for the angular velocities ω_{BC} and ω_{CD}. (Indicate CW or CCW in your answers.)

Eq. (a): $\xrightarrow{+}$

Eq. (b): $+\uparrow$

Solve Eqs. (a) and (b):

$\omega_{BC} =$ _____ $\omega_{CD} =$ _____ **Ans.**

Solution II *(using vector notation)*

(2) Write the relative velocity equation between B and C using vector notation.

The relative velocity equation is $\mathbf{v}_B = \mathbf{v}_C + \mathbf{v}_{B/C}$, which becomes

$$\boldsymbol{\omega}_{AB} \times \mathbf{r}_{B/A} = \boldsymbol{\omega}_{CD} \times \mathbf{r}_{C/D} + \boldsymbol{\omega}_{BC} \times \mathbf{r}_{B/C} \qquad \text{Eq. (a)}$$

Write the three angular velocities using vector notation:

$\boldsymbol{\omega}_{AB} =$ _____ ; $\boldsymbol{\omega}_{BC} =$ _____ ; $\boldsymbol{\omega}_{CD} =$ _____

Referring to the problem figure, write the three relative position vectors using vector notation:

$\mathbf{r}_{B/A} =$ _____ ; $\mathbf{r}_{C/D} =$ _____ ; $\mathbf{r}_{B/C} =$ _____

(continued)

In the space below, substitute the above terms into Eq. (a) and solve for the angular velocities. Express your answer using vector notation.)

$$\omega_{AB} \times r_{B/A} = \omega_{CD} \times r_{C/D} + \omega_{BC} \times r_{B/C}$$

$\omega_{BC} =$ _____ $\omega_{CD} =$ _____   **Ans.**

Lesson 21 Instant Centers for Velocity

Text Reference: Article 16.6; Sample Problems 16.7–16.9

A. SELF-TEST (*To be done after assigned reading has been completed.*)

Note: In this Self-Test, it is assumed that the body is undergoing plane motion.

1. Define "instant center for velocity".

   _____

2. When is it necessary to consider the "body extended"?

   _____

3. (a) The acceleration of the instant center of zero velocity is *always* zero. (T or F)____

   (b) The acceleration of the instant center of zero velocity is *never* zero. (T or F)____

4. When using the method of instant centers:
   (a) the velocity of any point in the body is _____ to the line drawn from the point to the instant center.
   (b) the magnitude of the velocity of any point in the body is proportional to the distance of the point from the instant center. (T or F) ____
   (c) the sense of the velocity vector of any point must be consistent with the sense of the _____ of the body.

B. GUIDED PROBLEMS

PROBLEM 21.1

Note: This problem is intended to give you practice in using instant centers for velocity.

The wheel shown is rolling without slipping on the horizontal plane. Using the fact that point C is the instant center, compute the magnitudes of the velocity of each of the labeled points, and show the direction of each velocity vector directly on the figure.

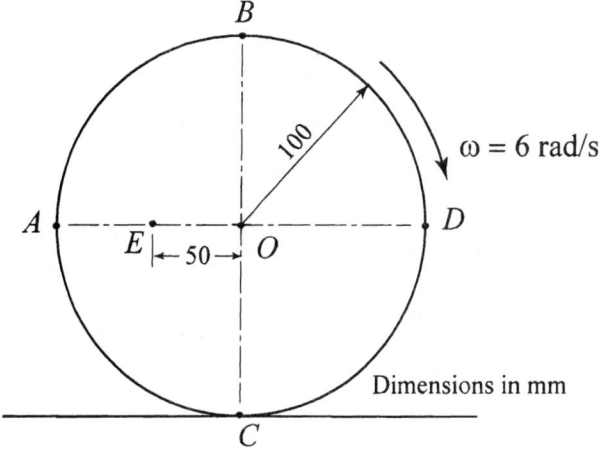

(a) $v_O =$ 600 mm/s

(b) $v_A =$ 848.5 mm/s

(c) $v_B =$ 1200 mm/s

(d) $v_D =$ 848.5 mm/s

(e) $v_E =$ 670.8 mm/s

PROBLEM 21.2

In the position shown, the angular velocity of bar AB is 4 rad/s clockwise. For this position, determine the angular velocities or bars AB and CD using the method of instant centers.

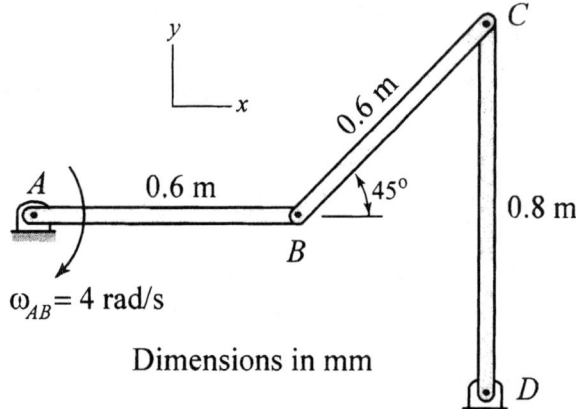

a. Comments and Analysis

- This problem was solved previously as Guided Problem 20.2 using relative velocity equations.

- Recall that if r is the distance from the instant center of a body to a point on the body, then the (magnitude) of the velocity of the point is $v = r\omega$, where ω is the angular velocity of the body. The direction of the velocity vector is determined by inspection.

- The steps that we will use in the analysis are:

    (1) Locate the instant center for bar AB; calculate the velocity of point B.

    (2) Locate the instant center for bar CD; determine the velocity of point C in terms of ω_{CD}.

    (3) Using the velocities of points B and C, locate the instant center for bar BC.

    (4) Using the instant center for bar BC, compute the angular velocities ω_{BC} and ω_{CD}.

b. Guided Solution

(1) Locate the instant center for bar AB; calculate the velocity of point B.

Because point A is a fixed point, it is the instant center for bar AB. Therefore, the velocity of B is: $v_B = r\omega_{AB} =$ _____ (directed up or down?)

(2) Locate the instant center for bar CD; determine the velocity of point C in terms of ω_{CD}.

Because point D is a fixed point, it is the instant center for bar CD. Therefore, assuming that ω_{CD} is counterclockwise, the velocity of C is

$$v_C = r\omega_{CD} = \underline{\qquad\qquad} \text{(directed left or right?)}$$

(*continued*)

(3) Using the velocities of points B and C, locate the instant center for bar BC.

The figure at the right shows body BC extended. On this figure, (i) draw the velocity vector for point B; (ii) draw the velocity vector for point C; (iii) locate the instant center of BC using the fact that a velocity vector is perpendicular to the line from the instant center. (Label this instant center as point O.)

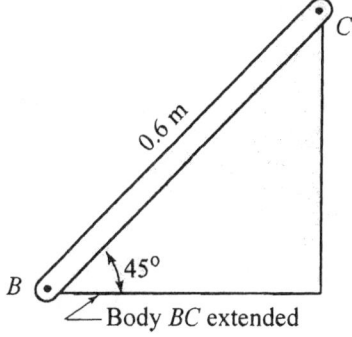

(4) In the space below, use point O to compute the angular velocities ω_{BC} and ω_{CD}. (Indicate CW or CCW in your answers.)

$\omega_{BC} =$ _____ ; $\omega_{CD} =$ _____ **Ans.**

A WORD OF CAUTION!!

The figure at the right shows a common "short-cut" method for locating point O. This figure can be misleading unless you are very careful. Note that the instant center O for bar BC appears to lie on bar CD, but it does not. The instant center O of body BC lies on body BC extended. At this instant, point O happens to be coincident with a point on bar CD. *Never assume that the instant center of one body lies on a different body.*

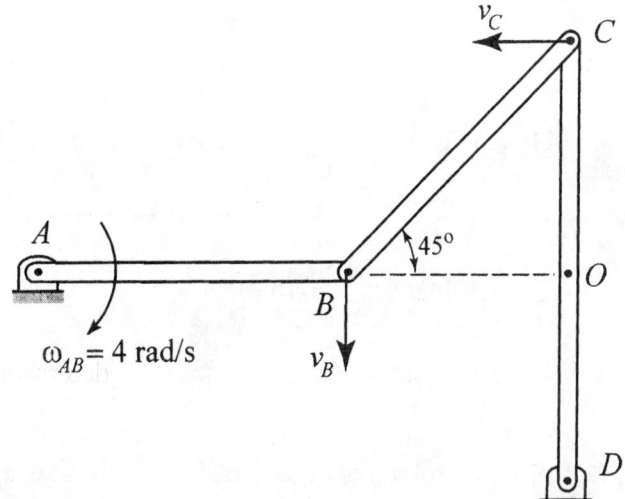

Lesson 22 Method of Relative Acceleration

Text Reference: Article 16.7; Sample Problems 16.10–16.12

A. SELF-TEST (*To be done after assigned reading has been completed.*)

Note: *In this Self-Test, it is assumed that the body is undergoing plane motion.*

1. Consider the relative acceleration equation between two arbitrary points A and B on the same rigid body undergoing plane motion: $\mathbf{a}_B = \mathbf{a}_A + \boldsymbol{\omega} \times (\boldsymbol{\omega} \times \mathbf{r}_{B/A}) + \boldsymbol{\alpha} \times \mathbf{r}_{B/A}$. Assuming that the relative position vector $\mathbf{r}_{B/A}$ is the only known quantity, what is the number of unknowns in each of the other terms:

    (a) $\mathbf{a}_B$: _____ ; (b) $\mathbf{a}_A$: _____ ; (c) $\boldsymbol{\omega}$: _____ ; (d) $\boldsymbol{\alpha}$: _____

    Therefore, the total number of unknowns in the equation is (e) _____
    How many variables must be known beforehand if the relative acceleration equation is to be solved? (f) _____ . Relative velocity analysis can usually be used to find which variable? (g) _____ .

2. A point for which the acceleration contains less than two unknowns is referred to as a _____ point for acceleration.

3. List the steps in the application of the relative acceleration method:

    Step 1: If the angular velocity of the body is unknown, find it by using the method of (a)_____.

    Step 2: Identify two (b)_____ points for acceleration.

    Step 3: Write the relative (c)_____ equation between the points.

    Step 4: Solve the equation if there are how many unknowns? (d) _____

88

B. GUIDED PROBLEM

PROBLEM 22.1

The bar AB is rotating clockwise with the constant angular velocity of 4 rad/s. In the position shown, determine the angular accelerations of bars BC and CD. Solve using scalar notation (Solution I) and vector notation (Solution II).

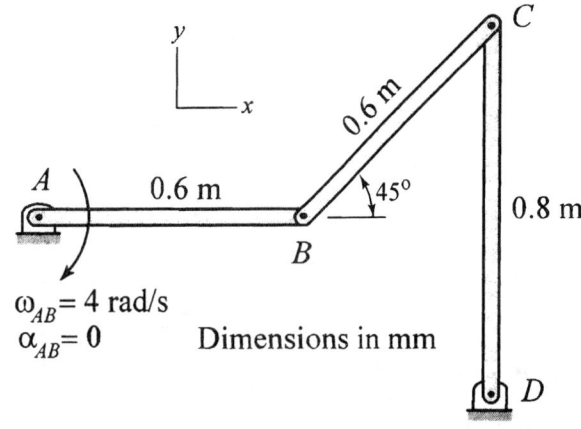

$\omega_{AB} = 4$ rad/s
$\alpha_{AB} = 0$ Dimensions in mm

a. Comments and Analysis

- Referring to the solutions to Guided Problems 20.2 and 21.2 we know that the angular velocities of bars BC and CD are $\omega_{BC} = 5.66$ rad/s CCW and $\omega_{CD} = 3.00$ rad/s CCW.

- The steps that we will use in the analysis are:

    (1) Identify the kinematically important points.

    (2) Write the relative acceleration equation between the appropriate kinematically important points, and solve for the equation for the angular accelerations α_{BC} and α_{CD}. (We will accomplish this step using both scalar and vector notation.)

b. Guided Solution

(1) Identify the kinematically important points.

As discussed in the solution to Guided Problem 20.2, the following are the kinematically important points: (i) the fixed points A and D; (ii) point B that travels on a circular path centered at A, and (iii) point C, that follows a circular path centered at D.

We will write the relative acceleration equation between points B and C which are kinematically important points that belong to the same rigid body.

(continued)

Solution I *(using scalar notation)*

(2) Complete the following relative acceleration equation between points B and C. Note that the directions of both α_{BC} and α_{CD} have been assumed to be counterclockwise.

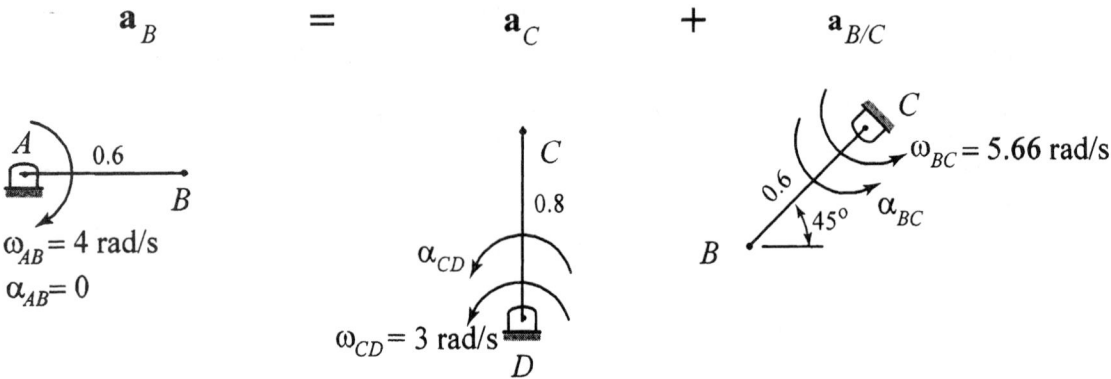

$$\mathbf{a}_B = \mathbf{a}_C + \mathbf{a}_{B/C}$$

In the space below, solve the above equation for the angular accelerations α_{BC} and α_{CD}. (Indicate CW or CCW in your answers.)

$+\uparrow$

$\xrightarrow{+}$

$\alpha_{CD} =$ _____ $\alpha_{BC} =$ _____ **Ans.**

Solution II *(using vector notation)*

(2) Write the relative acceleration equation between B and C using vector notation.

The relative acceleration equation is $\mathbf{a}_B = \mathbf{a}_C + \mathbf{a}_{B/C}$, which becomes

$$\boldsymbol{\omega}_{AB} \times (\boldsymbol{\omega}_{AB} \times \mathbf{r}_{B/A}) + \boldsymbol{\alpha}_{AB} \times \mathbf{r}_{B/A} = \boldsymbol{\omega}_{CD} \times (\boldsymbol{\omega}_{CD} \times \mathbf{r}_{C/D}) + \boldsymbol{\alpha}_{CD} \times \mathbf{r}_{C/D} + \boldsymbol{\omega}_{BC} \times (\boldsymbol{\omega}_{BC} \times \mathbf{r}_{B/C}) + \boldsymbol{\alpha}_{BC} \times \mathbf{r}_{B/C}$$

Write the three (known) angular velocities using vector notation:

$\boldsymbol{\omega}_{AB} =$ _____ ; $\boldsymbol{\omega}_{BC} =$ _____ ; $\boldsymbol{\omega}_{CD} =$ _____

(continued)

Write the three angular acceleration vectors (assume that α_{BC} and α_{CD} are counterclockwise) and the three relative position vectors (refer to the problem figure):

$\boldsymbol{\alpha}_{AB} =$ _____ ; $\boldsymbol{\alpha}_{BC} =$ _____ ; $\boldsymbol{\alpha}_{CD} =$ _____

$\mathbf{r}_{B/A} =$ _____ ; $\mathbf{r}_{C/D} =$ _____ ; $\mathbf{r}_{B/C} =$ _____

In the space below, write each term of the relative acceleration equation:

$\boldsymbol{\omega}_{AB} \times (\boldsymbol{\omega}_{AB} \times \mathbf{r}_{B/A}) =$

$\boldsymbol{\alpha}_{AB} \times \mathbf{r}_{B/A} =$

$\boldsymbol{\omega}_{CD} \times (\boldsymbol{\omega}_{CD} \times \mathbf{r}_{C/D}) =$

$\boldsymbol{\alpha}_{CD} \times \mathbf{r}_{C/D} =$

$\boldsymbol{\omega}_{BC} \times (\boldsymbol{\omega}_{BC} \times \mathbf{r}_{B/C}) =$

$\boldsymbol{\alpha}_{BC} \times \mathbf{r}_{B/C} =$

In the space below, substitute the above terms into the relative acceleration equation and solve for the angular accelerations. (Express your answers using vector notation.)

$\boldsymbol{\omega}_{AB} \times (\boldsymbol{\omega}_{AB} \times \mathbf{r}_{B/A}) + \boldsymbol{\alpha}_{AB} \times \mathbf{r}_{B/A} = \boldsymbol{\omega}_{CD} \times (\boldsymbol{\omega}_{CD} \times \mathbf{r}_{C/D}) + \boldsymbol{\alpha}_{CD} \times \mathbf{r}_{C/D} + \boldsymbol{\omega}_{BC} \times (\boldsymbol{\omega}_{BC} \times \mathbf{r}_{B/C}) + \boldsymbol{\alpha}_{BC} \times \mathbf{r}_{B/C}$

$\alpha_{BC} =$ _____ $\alpha_{CD} =$ _____   **Ans.**

Lesson 23 Absolute and Relative Derivatives of Vectors; Rotating Reference Frames

Text Reference: Articles 16.8 and 16.9; Sample Problems 16.13 and 16.14

Note: *In this Lesson, it is assumed that the body is undergoing plane motion.*

A. SELF-TEST (*To be done after assigned reading has been completed.*)

1. The *xyz* reference frame is fixed in space, the unit vectors along the axes denoted by **i**, **j**, and **k**.

   (a) Write a vector **V** in terms of the fixed axes:

   V =

   (b) The time derivative of this vector is its *absolute derivative*. (T or F) ____

2. The *x'y'z'* reference frame is embedded in a body **ß**, the unit vectors being **i'**, **j'**, and **k'**.

   (a) Write a vector **V** in terms of the embedded coordinates

   V =

   (b) Write the equation for the *relative derivative* (relative to the body **ß**) of **V**:

   $$\left(\frac{d\mathbf{V}}{dt}\right)_{/\beta} =$$

(continued)

3. (a) Complete the following relationship between the absolute derivative of a vector **V** and its derivative relative to the body ß, where ω is the angular velocity of ß:

$$\left(\frac{d\mathbf{V}}{dt}\right) = \left(\frac{d\mathbf{V}}{dt}\right)_{/\beta} + \underline{\hspace{2cm}}$$

(b) Write the above absolute derivative if **V** is embedded in the body ß:

$$\left(\frac{d\mathbf{V}}{dt}\right) = \underline{\hspace{3cm}}$$

(c) The absolute second derivative of **V** is

$$\frac{d^2\mathbf{V}}{dt^2} = \left(\frac{d^2\mathbf{V}}{dt^2}\right)_{/\beta} + \dot{\omega}\times\mathbf{V} + \omega\times(\omega\times\mathbf{V}) + 2\omega\times\left(\frac{d\mathbf{V}}{dt}\right)_{/\beta}$$

Write the above absolute second derivative if **V** is embedded in the body ß:

$$\frac{d^2\mathbf{V}}{dt^2} = \underline{\hspace{3cm}}$$

4. Consider the relative velocity equation: $\mathbf{v}_P = \mathbf{v}_A + \mathbf{v}_{P'/A} + \mathbf{v}_{P/\beta}$, where the $x'y'z'$ coordinate system is embedded in the body ß.

(a) Point P' moves independently of the body ß. (T or F) _____

(b) Point P moves independently of the body ß. (T or F) _____

(c) What is the relationship between points P' and P?
_____

(d) The term that represents the velocity of the origin of the $x'y'z'$ frame is _____

(e) Identify which term can be written as:

(i) $\left(\dfrac{d\mathbf{r}_{P/A}}{dt}\right)_{/\beta}$:

(ii) $\omega \times \mathbf{r}_{P/A}$:

(continued)

5. Consider the relative acceleration equation: $\mathbf{a}_P = \mathbf{a}_A + \mathbf{a}_{P'/A} + \mathbf{a}_{P/\beta} + \mathbf{a}_C$, where the $x'y'z'$ coordinate system is embedded in the body ß.

   (a) Point P' moves independently of the body ß. (T or F) _____

   (b) Point P moves independently of the body ß. (T or F) _____

   (c) The term that represents the acceleration of the origin of the $x'y'z'$ frame is _____

   (d) Write each of the following terms using $\mathbf{r}_{P/A}$ (the relative position vector of P relative to A), $\mathbf{v}_{P/\beta}$ (the velocity of P relative to the body ß), and the angular velocity $\boldsymbol{\omega}$ and angular acceleration $\dot{\boldsymbol{\omega}}$ of the body ß.

   (i) $\mathbf{a}_{P'/A} =$

   (ii) $\mathbf{a}_{P/\beta} =$

   (iii) $\mathbf{a}_C =$

B. GUIDED PROBLEMS

PROBLEM 23.1

The ends of bar AB remain in contact with the surfaces as the bar is falling down. In the position shown, the angular velocity of the bar is $\omega = 2$ rad/s CW, and the small collar P is sliding along the bar with the velocity relative to the bar of 5 ft/s, directed toward A. For this position, determine the velocity of P. Solve using scalar notation (Solution I) and vector notation (Solution II).

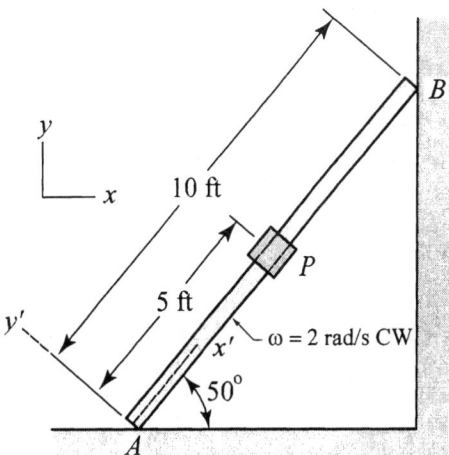

a. Comments and Analysis

- The following are the three points that will be used in our analysis:

    A: the origin of the $x'y'$-coordinate system that is embedded in bar AB

    P: the collar that is moving along bar AB

    P': the point (not shown) on the bar AB that is coincident with P at this instant

- The solution involves writing and solving the relative velocity equation

$$\mathbf{v}_P = \mathbf{v}_A + \mathbf{v}_{P'/A} + \mathbf{v}_{P/AB}$$

- This equation is equivalent to two scalar equations. Before attempting to solve the equation, we must be certain that it contains only two unknowns. Let us list the number of unknowns in each term in the above equation:

    $\mathbf{v}_P$ There are two unknowns because nothing is known about this velocity vector. We will label these unknowns using the Cartesian components of $\mathbf{v}_P$: $(v_P)_x$ and $(v_P)_y$.

    $\mathbf{v}_A$ This vector would, in general, contain two unknowns. However, performing a relative velocity analysis between points A and B will independently determine the velocity of point A. Therefore, there will be no unknowns.

    $\mathbf{v}_{P'/A}$ Both P' and A belong to the same rigid bar AB for which the angular velocity is known. Therefore, this term contains no unknowns.

    $\mathbf{v}_{P/AB}$ There are no unknowns because the velocity of P relative to the bar AB is given.

Observe that the total number of unknowns will be two, $(v_P)_x$ and $(v_P)_y$, after the velocity of point A has been determined by relative velocity analysis.

(continued)

a. Comments and Analysis *(continued)*

- The steps that we will use in the analysis are:

    (1) Compute $\mathbf{v}_A$ by performing a relative velocity analysis between A and B.

    (2) Write and solve the relative velocity equation $\mathbf{v}_P = \mathbf{v}_A + \mathbf{v}_{P'/A} + \mathbf{v}_{P/AB}$.

    (We will accomplish this step using both scalar and vector notation.)

b. Guided Solution

(1) Compute $\mathbf{v}_A$ by performing a relative velocity analysis between A and B.

The velocity of point A can be found using the relative velocity equation $\mathbf{v}_A = \mathbf{v}_B + \mathbf{v}_{A/B}$, or by using the method of instant centers. Omitting the details, the velocity of point A in the position shown is 15.32 ft/s, directed to the left.

Solution I *(using scalar notation)*

(2) Write the relative velocity equation using scalar notation.

Complete the following relative velocity equation by writing each term in the dotted boxes using scalar notation.

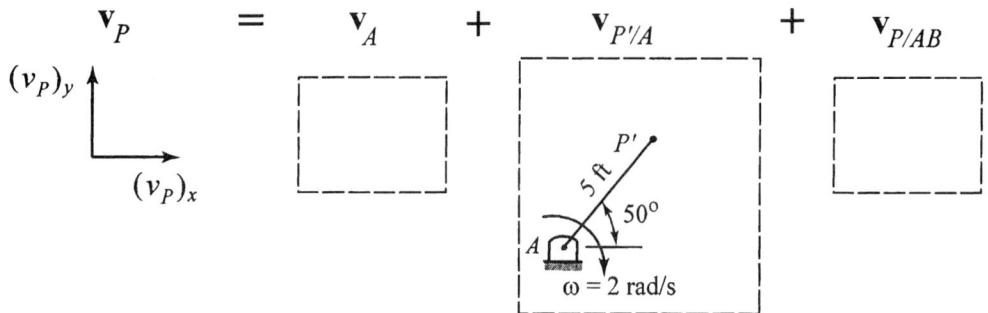

In the space below, solve the above equation for $(v_P)_x$ and $(v_P)_y$. Determine the velocity of the collar P. Express your answer using vector notation.

$+\rightarrow$

$+\uparrow$

$\mathbf{v}_P = $ _____ **Ans.**

(continued)

Solution II *(using vector notation)*

(2) Using vector notation, write each of the terms on the right side of the relative velocity equation $\mathbf{v}_P = \mathbf{v}_A + \mathbf{v}_{P'/A} + \mathbf{v}_{P/AB}$.

$\mathbf{v}_A =$

$\boldsymbol{\omega} =$ $\qquad$ $\mathbf{r}_{P'/A} =$

$\mathbf{v}_{P'/A} = \boldsymbol{\omega} \times \mathbf{r}_{P'/A} =$

$\mathbf{v}_{P/AB} =$

Substitute the above terms into the relative velocity equation and compute the velocity of the collar *P*.

$\mathbf{v}_P =$

$\mathbf{v}_P =$ _____ **Ans.**

- -

PROBLEM 23.2

The ends of bar AB remain in contact with the surfaces as the bar is falling down. In the position shown, the angular velocity of the bar is $\omega = 2$ rad/s CW, and its angular acceleration is $\alpha = 3$ rad/s^2 CCW. The small collar P is sliding along the bar with the velocity and acceleration relative to the bar of 5 ft/s and 10 ft/s^2, respectively, both directed toward A. For this position, determine the acceleration of P. Solve using scalar notation and vector notation

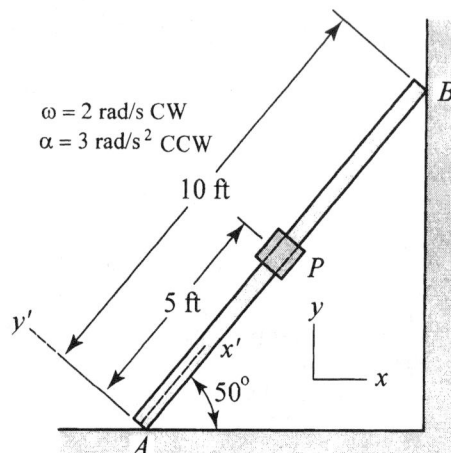

a. Comments and Analysis

- The following are the three points that will be used in our analysis:

    A: the origin of the $x'y'$-coordinate system that is embedded in bar AB

    P: the collar that is moving along bar AB

    P': the point (not shown) on the bar AB that is coincident with P at this instant

- The solution involves writing and solving the relative acceleration equation

    $$\mathbf{a}_P = \mathbf{a}_A + \mathbf{a}_{P'/A} + \mathbf{a}_{P/AB} + \mathbf{a}_C \quad (\mathbf{a}_C \text{ is the Coriolis acceleration})$$

- This equation is equivalent to two scalar equations. Before attempting to solve the equation, we must be certain that it contains only two unknowns. Let us list the number of unknowns in each term in the above equation:

    $\mathbf{a}_P$ There are two unknowns because nothing is known about this acceleration vector. We will label these unknowns using the Cartesian components of $\mathbf{a}_P$: $(a_P)_x$ and $(a_P)_y$.

    $\mathbf{a}_A$ This vector would, in general, contain two unknowns. However, performing a relative acceleration analysis between points A and B will independently determine the acceleration of point A. Therefore, there will be no unknowns.

    $\mathbf{a}_{P'/A}$ Both P' and A belong to the same rigid bar AB the angular velocity and angular acceleration are known. Therefore, this term contains no unknowns.

    $\mathbf{a}_{P/AB}$ There are no unknowns because the acceleration of P relative to the bar AB is given.

    $\mathbf{a}_C$ There are no unknowns because the angular velocity ω and $\mathbf{v}_{P/AB}$ are given.

(continued)

a. **Comments and Analysis** *(continued)*

Observe that the total number of unknowns will be two, $(a_P)_x$ and $(a_P)_y$, after the acceleration of point A has been determined by relative acceleration analysis.

- The steps that we will use in the analysis are:

    (1) Compute $\mathbf{a}_A$ by performing a relative acceleration analysis between A and B.

    (2) Write and solve the relative acceleration equation $\mathbf{a}_P = \mathbf{a}_A + \mathbf{a}_{P'/A} + \mathbf{a}_{P/AB} + \mathbf{a}_C$.

    (We will accomplish this step using both scalar and vector notation.)

b. **Guided Solution**

(1) Compute $\mathbf{a}_A$ by performing a relative acceleration analysis between A and B.

The acceleration of point A can be found using the relative acceleration equation $\mathbf{a}_A = \mathbf{a}_B + \mathbf{a}_{A/B}$. Omitting the details, the acceleration of point A in the position shown is 48.69 ft/s², directed to the right.

Solution I *(using scalar notation)*

(2) Write the relative acceleration vector using scalar notation.

Complete the following relative acceleration equation by writing each term in the dotted boxes using scalar notation (units ft/s²).

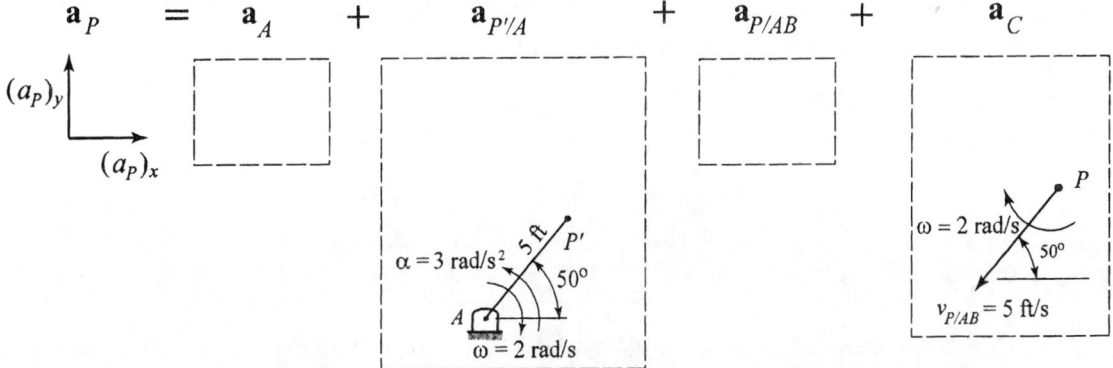

In the space below, solve the above equation for $(a_P)_x$ and $(a_P)_y$. Determine the acceleration of the collar P. Express your answer using vector notation.

$\overset{+}{\longrightarrow}$

$+\uparrow$

$\mathbf{a}_P = $ _____ **Ans**

(continued)

Solution II *(using vector notation)*

(2) Using vector notation, write each of the terms on the right side of the relative acceleration equation $\mathbf{a}_P = \mathbf{a}_A + \mathbf{a}_{P'/A} + \mathbf{a}_{P/AB} + \mathbf{a}_C$.

$\mathbf{a}_A =$

$\boldsymbol{\omega} =$ $\qquad$ $\boldsymbol{\alpha} =$ $\qquad$ $\mathbf{r}_{P'/A} =$

$\mathbf{a}_{P'/A} = \boldsymbol{\omega} \times (\boldsymbol{\omega} \times \mathbf{r}_{P'/A}) + \boldsymbol{\alpha} \times \mathbf{r}_{P'/A}$

$\mathbf{a}_{P'/A} =$

$\mathbf{a}_{P/AB} =$

$\mathbf{a}_C = 2\boldsymbol{\omega} \times \mathbf{v}_{P/AB} =$

Substitute the above terms into the relative acceleration equation and compute the acceleration of the collar P in the space below.

$\mathbf{a}_P =$

$\mathbf{a}_P =$ _____ **Ans**

Lesson 24 Mass Moment of Inertia; Composite Bodies

Text Reference: Articles 17.1 and 17.2; Sample Problems 17.1 and 17.2

A. SELF-TEST *(To be done after assigned reading has been completed.)*

1. A body of mass m occupies a region V. Letting r be the perpendicular distance from the a-axis to the differential mass dm, write the defining equation for the moment of inertia of the body about the a-axis:

    $I_a =$

2. The moment of inertia of a body of mass m about the a-axis is I_a. Write the defining equation for the radius of gyration of the body about the a-axis:

    $k_a =$

3. The radius of gyration is a physical measurement. (T or F) _____

4. What is meant by the *central a*-axis?

    _____

5. In words, define each of the terms that appear in the parallel-axis theorem:

    $$I_a = \bar{I}_a + md^2$$

    (a) I_a: _____

    (b) $\bar{I}_a$: _____

    (c) m: _____

    (d) d: _____

6. The method of composite bodies follows from which property of definite integrals?

    _____

B. GUIDED PROBLEM

PROBLEM 24.1

The assembly consists of three homogenous slender bars that are connected together. Compute the mass moment of inertia of the assembly about the *x*-axis.

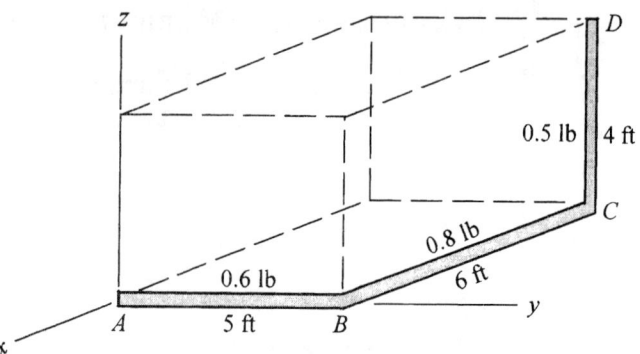

a. Comments and Analysis

- Referring to the tables in your textbook, the inertial properties of the slender bar shown at the right are:

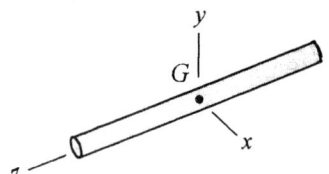

$$\bar{I}_x = \bar{I}_y = \frac{mL^2}{12} \qquad \bar{I}_z \approx 0$$

- The mass moment of inertia of the assembly about the *x*-axis equals the sum of the mass moments of inertia about that axis for the three bars that form the assembly.

- Because the *x*-axis for the assembly is not the central axis for any of the bars, our solution will involve using the parallel-axis theorem for each bar.

- The parallel-axis theorem is $I_a = \bar{I}_a + md^2$, where m is the mass of the body, I_a is the moment of inertia of the body about the *a*-axis, $\bar{I}_a$ is the moment of inertia of the body about the central *a*-axis, and d is the distance between the two axes.

- The steps that we will use in the computation are:

   (1) For each bar: (i) calculate $\bar{I}_x$; (ii) refer to the problem figure to determine the distance d between the central *x*-axis and the *x*-axis, and
   (iii) use the parallel-axis theorem to calculate I_x.
   (2) Sum the moments of inertia about the *x*-axis for the three bars.

(continued)

b. **Guided Solution**

(1) Calculate $\bar{I}_x$ and then I_x for each bar.

Bar AB

$\bar{I}_x =$

$d =$

$I_x = \bar{I}_x + md^2 =$

Bar BC

$\bar{I}_x =$

$d =$

$I_x = \bar{I}_x + md^2 =$

Bar CD

$\bar{I}_x =$

$d =$

$I_x = \bar{I}_x + md^2 =$

(2) Calculate I_x for the assembly.

$(I_x)_{\text{assembly}} = \Sigma(I_x)$

$(I_x)_{\text{assembly}} = $ _____ **Ans.**

Lesson 25 Angular Momentum of a Rigid Body, Equations of Motion; Force-Mass-Acceleration Method

Text Reference: Articles 17.3–17.5; Sample Problems 17.3–17.9

Note: The angular momentum equations presented in Art. 17.3 were used exclusively to derive the equations of motion for a rigid body in Art. 17.4. The angular momentum equations will be used in the impulse-momentum method of analysis in Chapter 18. Therefore, this Lesson focuses only upon application of the equations of motion to the plane motion of a rigid body.

A.1 SELF-TEST *(To be done after assigned reading has been completed.)*

1. Identify each of the terms in the equations of motion for a rigid body: $\Sigma \mathbf{F} = m\bar{\mathbf{a}}$.

    (a) $\Sigma \mathbf{F}$: _____

    (b) m: _____

    (c) $\bar{\mathbf{a}}$: _____

2. The moment equation for the plane motion of a rigid body of mass m can be written as $\Sigma M_A = \bar{I}\alpha + m\bar{a}d$, where A is an arbitrary point. Identify each of the following terms:

    (a) The term $\bar{I}$ refers to _____

    (b) The term α refers to _____

    (c) The term $\bar{a}$ refers to _____

    (d) The term d refers to _____

3. Consider the special case of the moment equation for a rigid body: $\Sigma M_A = I_A \alpha$. Assuming that point A is not the mass center, what are the conditions on the choice of point A for this equation to be valid?

    _____

Note: Assuming that the moment equation $\Sigma M_A = I_A \alpha$ is valid for *any* point A is a a very common mistake.

(continued)

4. Write the moment equation if the moment center is chosen to be the mass center G of a rigid body: $\Sigma M_G =$

5. In general, the number of independent equations of motion for the plane motion of a rigid body is _____.

6. The mass-acceleration diagram (MAD) consists of the vector $m\bar{a}$, called the (a) _____, and the couple $\bar{I}\alpha$, called the (b) _____. On the MAD the vector $m\bar{a}$ is shown acting at (c) _____.

7. The equations of motion for a rigid body can be determined using the equivalence of the FBD and the MAD. (T or F) ____

8. Describe the MAD for a translating rigid body.
   _____

9. Describe the MAD for a body that is rotating about its mass center:
   _____

A.2 GENERAL COMMENT

You will notice that there are seven sample problems that follow the reading assignment for this lesson. These problems can be classified as follows:

   (i) translation (Sample Problem 17.3)
   (ii) rotation about a fixed point (Sample Problems 17.4–17.6)
   (iii) general plane motion (Sample Problems 17.7–17.9)

Although there are three different classifications, be sure to recognize that the method of analysis for each problem is the same:

   (1) **draw** the FBD
   (2) **analyze** the kinematics
   (3) **draw** the MAD
   (4) **derive and solve** the three equations of motion

B. GUIDED PROBLEMS

PROBLEM 25.1

The uniform slender bar is 10 ft long and weighs 20 lb. The ends A and B of the bar are connected to collars that slide on rails that lie in the vertical plane. The bar is sliding to the right under the action of the force $P = 8$ lb. Determine $\bar{a}$, the acceleration of the mass center of the bar, and the forces acting on the collar at A and B. Neglect friction and weights of the collars.

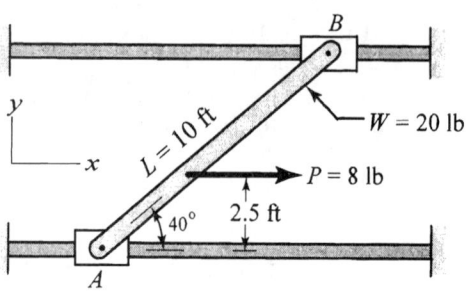

a. Comments and Analysis

- It is convenient to draw the free-body diagram (FBD) and mass-acceleration diagram (MAD) of the assembly consisting of the bar and the collars. (We would remove the collar from these diagrams only if we were interested in the pin reactions acting between the collars and the bar.)

- In general plane motion, the MAD would contain three terms: the inertia couple $\bar{I}\alpha$, and two scalar components of the inertia vector vector $m\bar{\mathbf{a}}$. In this problem, the bar is translating in the x-direction. Therefore, $\bar{I}\alpha$ ($\alpha = 0$) and $m\bar{a}_y$ will not appear on the MAD.

- The steps that we will use in the analysis are:
  (1) Draw the FBD and MAD of the assembly.
  (2) Determine the total number of unknowns on the FBD and MAD. Compare this number with the number of independent equations of motion.
  (3) Write and solve the equations of motion to determine the requested unknowns.

(continued)

b. Guided Solution

(1) Draw the FBD and MAD for the assembly using the figures below. Label the reactions at A and B as N_A and N_B, both assumed to act upward. Assume the inertia vector (magnitude $\bar{a}$) is directed to the right. Calculate the distances a, b, and c.

$a = $ _____

$b = $ _____

$c = $ _____

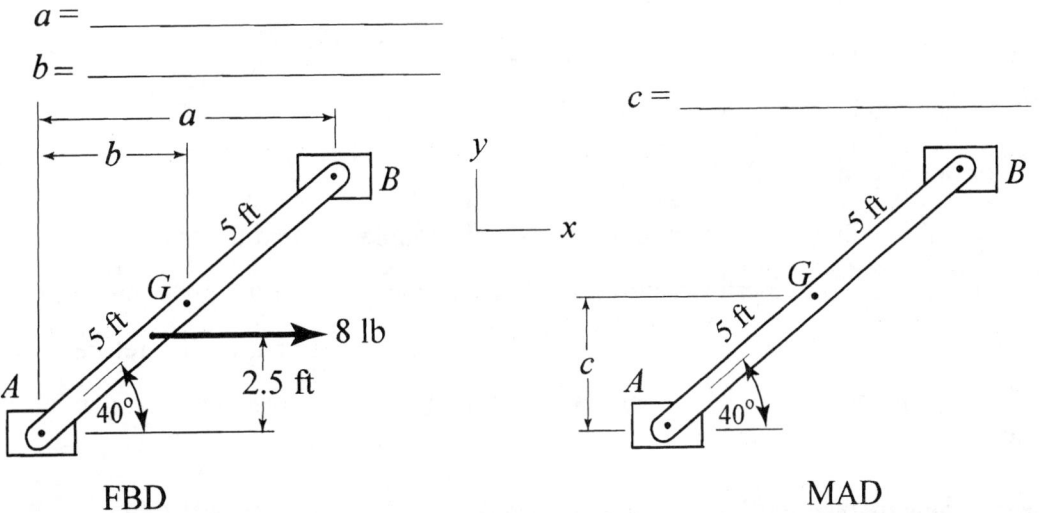

FBD   MAD

(2) The number of unknowns on the FBD is _____. The number of unknowns on the MAD is _____. Therefore, the total number of unknowns is _____. The total number of independent equations of motion is _____.

(3) In the space below, compute the N_A, N_B, and $\bar{a}$ using the equations indicated:

$\Sigma F_x = m\bar{a}_x$

$\xrightarrow{+}$

$\Sigma (M_A)_{FBD} = \Sigma(M_A)_{MAD}$

$\circlearrowleft +$

$\Sigma F_y = m\bar{a}_y$

$+\uparrow$

$N_A = $ _____ $N_B = $ _____ $\bar{a} = $ _____ **Ans.**

PROBLEM 25.2

The uniform thin hoop, with mass $m = 6$ kg and radius $R = 1.2$ m, rotates in a vertical plane about the pin O. In the position shown, the angular velocity of the hoop is $\omega = 3$ rad/s, clockwise. For this position, compute the angular acceleration α of the hoop, and R_O, the magnitude of the pin reaction at O.

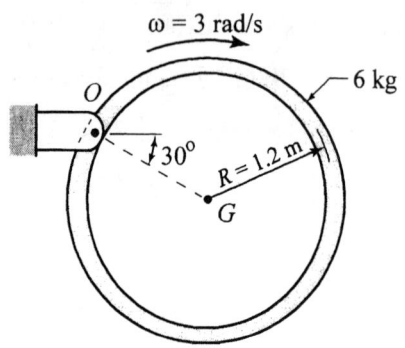

a. Comments and Analysis

- The MAD will contain three terms: the inertia couple $\bar{I}\alpha$, and the two scalar components of the inertia vector $m\bar{\mathbf{a}}$. Because the hoop is rotating about a fixed point, the path of the mass center G will be a circle centered at O. Therefore, it will be convenient to describe the inertia vector in terms of its normal and tangential components.

- The steps that we will use in the analysis are:

    (1) Draw the FBD of the hoop.

    (2) Draw MAD for the hoop.

    (3) Determine the total number of unknowns on the FBD and MAD. Compare this number with the number of independent equations of motion.

    (4) Write and solve the equations of motion to determine α and R_O.

b. Guided Solution

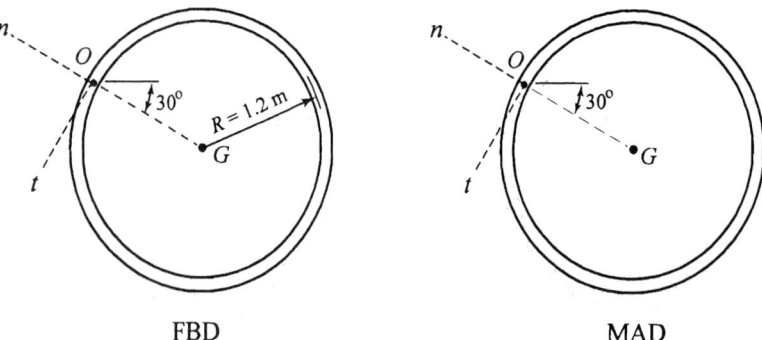

FBD　　　　　　　　　MAD

(1) On the figure above, draw the FBD of the hoop. Show the components of the pin reaction at O as the normal-tangential components, O_n and O_t (assume that each component acts in the positive coordinate direction shown).

(continued)

(2) *Preliminary computations required for the MAD*:

    Calculate $\bar{I}$ for the hoop: $\bar{I} =$

    Calculate the normal and tangential components of the inertia vector:

$$ma_n = mr\omega^2 =$$

$$ma_t = mr\alpha =$$

Using these values, draw the MAD for the hoop on the above figure (assume α is clockwise).

(3) The number of unknowns on the FBD is _____. The number of unknowns on the MAD is _____. Therefore, the total number of unknowns is _____. The total number of independent equation of motion is _____.

(4) In the space below, compute O_n, O_t, and α using the equations indicated:

$\Sigma(M_O)_{FBD} = \Sigma(M_O)_{MAD}$

$(+\circlearrowleft)$

Note: The moment equation $\Sigma(M_O)_{FBD} = I_O \alpha$ will yield the same result for α.

$\Sigma F_n = m\bar{a}_n$

$\nwarrow +$

$\Sigma F_t = m\bar{a}_t$

$+\swarrow$

Compute the magnitude of the pin reaction at O: $R_O =$

                         $\alpha =$ _____ ; $R_O =$ _____ **Ans.**

PROBLEM 25.3

The uniform slender 8-ft bar weighs 30 lb. The bar is released from rest in the position shown. For this position, compute the angular acceleration α of the bar, the acceleration a_A of the roller at A, and N_A, the reaction at A. Neglect the weight of the roller.

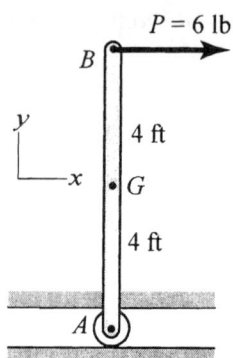

a. Comments and Analysis

- The only kinematic constraint is the path of A is a horizontal straight line. Therefore, the bar will undergo general plane motion.

- For general plane motion, the MAD will consist of three terms: the inertia couple $\bar{I}\alpha$ and the two scalar components of the inertia vector $m\bar{\mathbf{a}}$. The solution to this problem will involve kinematic analysis to relate the acceleration of the mass center G to the acceleration of the kinematically important point A.

- Because the bar is released from rest, there is no angular velocity in the position shown.

- The steps that we will use in the analysis are:

  (1) Draw the FBD of bar.

  (2) Write the relative acceleration equation between G and A to find $\bar{\mathbf{a}}$ in terms of the angular acceleration α of the bar.

  (3) Calculate $\bar{I}$ for the bar and draw its MAD.

  (4) Determine the total number of unknowns on the FBD and MAD. Compare this number with the number of independent equations of motion.

  (5) Write and solve the equations of motion to determine α, a_A, and N_A.

(continued)

b. Guided Solution

(1) Draw the FBD on the figure below.

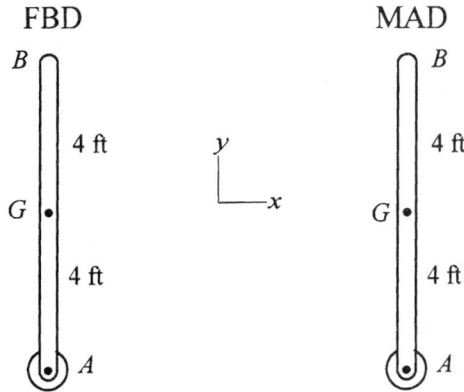

(2) Complete the following relative acceleration equation between points G and A. Assume that α is CW and that a_A is directed to the right. (The equation below is written using scalar notation. The equivalent equation using vector notation could have been used.)

$$\bar{\mathbf{a}} = \mathbf{a}_G = \mathbf{a}_A + \mathbf{a}_{G/A}$$

$$\underset{\omega = 0}{\underset{4 \text{ ft}}{}}$$

Therefore the kinematic relationship between a_G and α is $a_G = $

(3) Calculate $\bar{I}$ for the bar: $\bar{I} = \dfrac{mL^2}{12} =$

Using $\bar{I}$ and the results of the relative acceleration equation in Step (2), draw the MAD for the bar in the figure in Step (1).

(4) The number of unknowns on the FBD is _____. The number of unknowns on the MAD is _____. Therefore, the total number of unknowns is _____. The total number of independent equation of motion is _____.

(*continued*)

(5) In the space below, compute α, a_A, and N_A using the equations indicated:

$\Sigma M_G = \bar{I}\alpha$

$\Sigma F_x = m\bar{a}_x$

$\xrightarrow{+}$

$\Sigma F_y = m\bar{a}_y$

$+\uparrow$

$\alpha = $ _____ ; $a_A = $ _____ ; $N_A = $ _____ **Ans.**

Lesson 26 Work, Power of a Couple; Kinetic Energy of a Rigid Body

Text Reference: Articles 18.1-18.3; Sample Problems 18.1–18.3

A. SELF-TEST (*To be done after assigned reading has been completed.*)

1. The work done by a couple is derived by computing the work done by its two forces. (T or F) ___

2. (a) Write the equation for the work done by a couple C of variable magnitude for plane motion, where θ_1 and θ_2 are the initial and final angular positions measured from a convenient reference line.

    $U_{1\text{-}2} =$

   (b) Write the equation for the work done if the magnitude of the couple C remains constant during the angular rotation from θ_1 to θ_2.

    $U_{1\text{-}2} =$

3. (a) A 10-N·m constant clockwise couple rotates through a clockwise angle of 30°. Calculate the work done by the couple.

    $U_{1\text{-}2} =$

   (b) Calculate the work done by the couple in part (a) if the rotation is counterclockwise.

    $U_{1\text{-}2} =$

4. Calculate the power P in watts for a 10-N·m constant clockwise couple that is rotating counterclockwise with an angular velocity of 20 rad/s.

    $P =$

(continued)

5. The kinetic energy T of a rigid body undergoing plane motion can be written as

$$T = \frac{1}{2}m\bar{v}^2 + \frac{1}{2}\bar{I}\omega^2$$

   Identify each of the following terms:

   (a) m: _____

   (b) $\bar{v}$: _____

   (c) $\bar{I}$: _____

   (d) ω: _____

6. The kinetic energy of a rigid body in plane motion can be calculating using the equation $T = \frac{1}{2}I_A\omega^2$. What restriction must be placed on the choice of point A for this equation to be valid? _____

B. GUIDED PROBLEMS

PROBLEM 26.1

The two 8-lb blocks A and B hang from cables that are wrapped around the 20-lb homogeneous pulley. The pulley is rotating about a pin at its center of gravity G. If the constant clockwise couple $C = 1.2$ lb·ft is applied directly to the pulley, determine the work done on the system during one clockwise revolution of the pulley.

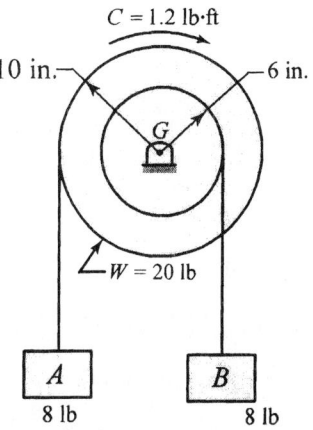

a. **Comments and Analysis**

- The total work done on the system is the sum of the work done by the weights of the blocks and the couple C. The weight of the pulley does not do work because the center of gravity does not move, the cable tensions are internal workless forces, and the pin reaction is an external workless force.

b. **Guided Solution**

(1) Calculate each of the following items for a clockwise rotation of one revolution

   (a) Couple C

   The clockwise angle turned through (in radians): $\Delta\theta =$

   $(U_{1-2})_C =$

   (b) Weight of A

   The distance d_A moved by the weight of A:

   $d_A =$ \hspace{2cm} (does A move up or down?)

   $(U_{1-2})_{W_A} =$

   (c) Weight of B

   The distance d_B moved by the weight of B:

   $d_B =$ \hspace{2cm} (does B move up or down?)

   $(U_{1-2})_{W_B} =$

(2) Compute the total work done:

   $(U_{1-2})_{total} =$ \underline{\hspace{6cm}} **Ans.**

- -

PROBLEM 26.2

In the position shown, the homogeneous cylinder is rolling without slipping with the clockwise angular velocity $\omega_{cyl} = 4$ rad/s. The weights of the block A, the uniform

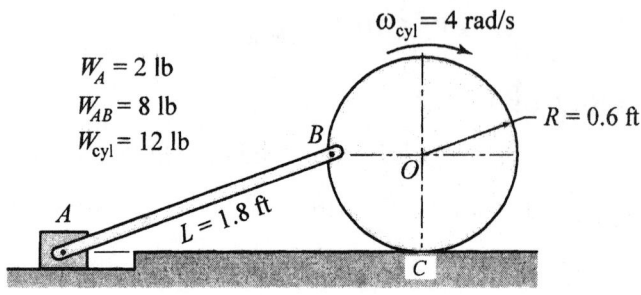

slender bar AB, and the cylinder are shown in the figure. Determine the total kinetic energy of the system in this position.

a. Comments and Analysis

- The steps that we will use in the analysis are:

    (1) Using relative velocity analysis, compute the velocity v_A of the block A and the angular velocity ω_{AB} of bar AB. (We choose to use the method of instant centers.)

    (2) Calculate the kinetic energy of each part of the system.

    (3) Sum the kinetic energies of the parts to obtain the total kinetic energy of the system.

b. Guided Solution

(1) Compute v_A and ω_{AB} using the method of instant centers.

We begin by noting that point C is the instant center of the cylinder and that $\mathbf{v}_A$ is horizontal. The figure below shows the geometric construction that was required to determine the location of point E, the instant center for bar AB:

   (i) $\mathbf{v}_B$ is perpendicular to BC because B is a point on the cylinder

   (ii) EB was drawn perpendicular to $\mathbf{v}_B$

   (iii) EA was drawn perpendicular to $\mathbf{v}_A$

   (iv) After locating point E, trigonometry was used to determine the distances shown in the figure (G is the center of gravity of bar AB)

(continued)

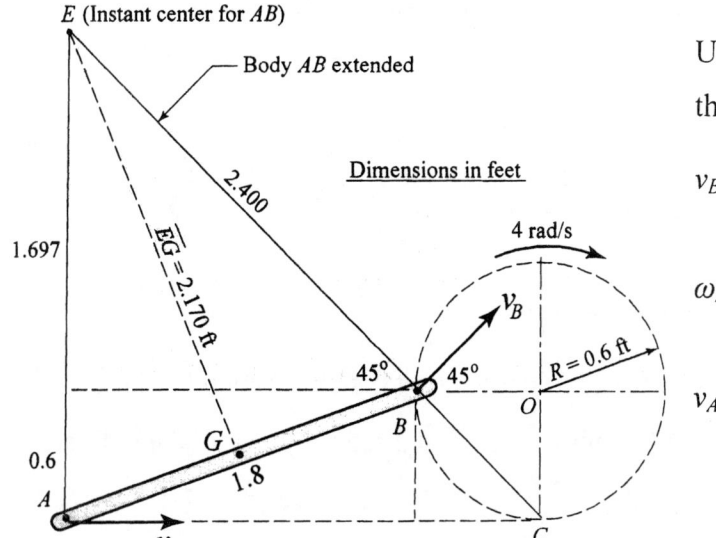

Using instant centers, calculate the following velocities

$v_B =$

$\omega_{AB} =$

$v_A =$

(2) Calculate the kinetic energy of each part. (Refer to the tables for inertial properties.)

Cylinder

$$I_C = \bar{I} + md^2 =$$

$$T_{cyl} = \frac{1}{2} I_C \omega_{cyl}^2 =$$

Bar AB

$$I_E = \bar{I} + md^2 =$$

$$T_{AB} = \frac{1}{2} I_E \omega_{AB}^2 =$$

Block A

$$T_A = \frac{1}{2} m_A v_A^2 =$$

(3) The total kinetic energy of the system becomes:

$$T_{total} = T_{cyl} + T_{AB} + T_A =$$

$T_{total} =$ _____ **Ans.**

Lesson 27 Work-Energy Principle and Conservation of Mechanical Energy

Text Reference: Article 18.4; Sample Problems 18.4–18.6

A. SELF-TEST (*To be done after assigned reading has been completed.*)

1. Identify each term in the work-energy principle for a system of connected rigid bodies: $(U_{1-2})_{ext} + (U_{1-2})_{int} = \Delta T$. (Subscripts 1 and 2 refer to the initial and final positions of the system, respectively.)

    (a) $(U_{1-2})_{ext}$: _____

    (b) $(U_{1-2})_{int}$: _____

    (c) ΔT: _____

2. Indicate whether each of the following is "workless" or "capable of doing work" on a system of rigid bodies when it is *internal* to the system.

    (a) Smooth pin connection: _____

    (b) Spring: _____

    (c) Friction force: _____

    (d) Inextensible string: _____

3. For the mechanical energy of a system of connected rigid bodies to be conserved, both internal and external forces must be conservative. (T or F) ____

B. GUIDED PROBLEM

PROBLEM 27.1

The 10-kg slender bar, 500 mm long, is attached to rollers A and B. The unstretched length of the spring is 350 mm and the spring constant is $k = 800$ N/m. The system is released from rest in Position 1, where the length of the spring is $L_1 = 500$ mm. Determine the angular velocity ω_2 of the bar when it reaches the vertical position shown in Position 2. Neglect friction and the masses of the rollers.

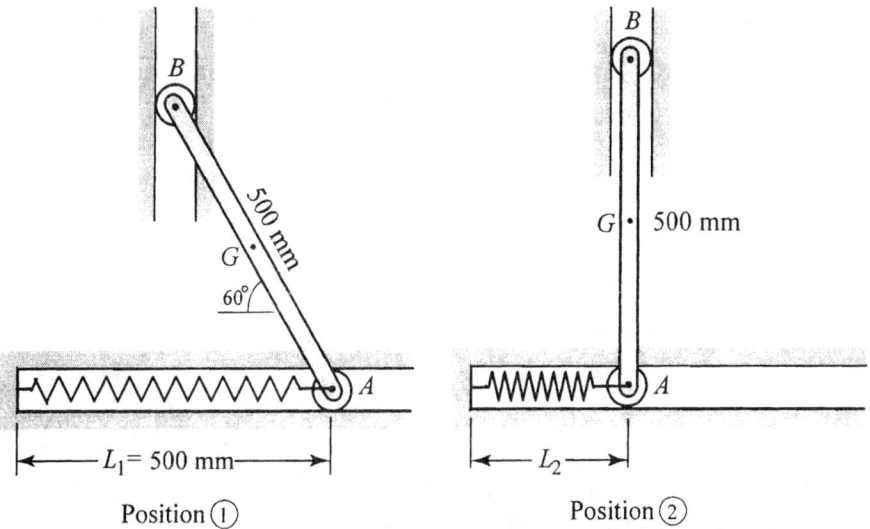

a. **Comments and Analysis**

- This problem is well-suited for solution by the work-energy method because it involves the change in velocity that occurs during a change in position. (The conservation of mechanical energy could also be used.)
- The steps that we will use in the analysis are:

    (1) Calculate work done by the spring between Positions 1 and 2.

    (2) Calculate work done by the weight W of the bar between Positions 1 and 2.

    (3) Write the expression for the kinetic energy of the bar in the Positions 1 and 2.

    (4) Write the work-energy principle between Positions 1 and 2, and solve for the final angular velocity ω_2 of the bar.

(continued)

b. Guided Solution

(1) Calculate the work done by the spring between Positions 1 and 2.

$$\text{Unstretched length of the spring: } L_0 =$$

Position 1: Length of spring: $L_1 =$ Deformation: $\delta_1 = L_1 - L_0 =$

Position 2: Length of spring: $L_2 =$

$$\text{Deformation: } \delta_2 = L_2 - L_0 =$$

The work done by the spring between Positions 1 and 2 is:

$$(U_{1-2})_{\text{spring}} = -\frac{1}{2}k(\delta_2^2 - \delta_1^2) =$$

(2) Calculate the work done by the weight W of the bar between Positions 1 and 2.

The vertical distance *upward* that G moves between Positions 1 and 2 is:

$\Delta h =$

The work done by the weight W between Positions 1 and 2 is:

$$(U_{1-2})_W = -W\Delta h =$$

(3) Write the expression for the kinetic energy of the bar in Positions 1 and 2.

Position 1: $T_1 =$

Position 2: The instant center for the bar in position 2 is located at point ____ .

The moment of inertia about that instant center is (refer to tables):

$I_{\text{i.c.}} =$

Therefore, the kinetic energy in position 2 is

$$T_2 = \frac{1}{2} I_{\text{i.c.}} \omega_2^2 =$$

(4) Write the work-energy principle between Positions 1 and 2, and solve for the final angular velocity ω_2 of the bar: $(U_{1-2})_{\text{spring}} + (U_{1-2})_W = T_2 - T_1$

$\omega_2 =$ _____   **Ans.**

Lesson 28 Momentum Diagrams; Impulse-Momentum Principles

Text Reference: Article 18.5 and 18.6; Sample Problems 18.7–18.11

Note: The reading assignment brings together topics that you have studied previously.

Article 18.5: Reviews the concept of angular momentum of a rigid body (Art. 17.3).

Article 18.6: Applies impulse-momentum for systems of particles (Arts. 14.6 and 15.7) to rigid body motion.

A. SELF-TEST *(To be done after assigned reading has been completed.)*

1. The angular momentum of a rigid body about an arbitrary point can be obtained from the momentum diagram of the body. (T or F) _____

2. The momentum diagram for a rigid body consists of the following two items: the linear momentum vector $\mathbf{p} = m\bar{\mathbf{v}}$ acting at the mass center G, and the angular momentum (couple) $h_G = \bar{I}\omega$.

    Identify each of the following terms:

    (a) m: _____

    (b) $\bar{\mathbf{v}}$: _____

    (c) $\bar{I}$: _____

    (d) ω: _____

3. If point A is the instant center for velocity, the angular momentum about A can be calculated using the equation $h_A = I_A \omega$. (T or F) _____

4. A homogeneous slender rod is rotating clockwise about one of its ends. The angular momentum about all points on the bar is also clockwise. (T or F) _____

5. The point about which the angular momentum of a rotating rigid body is zero is called the _____

(continued)

6. Consider the linear impulse-momentum equation for a rigid body, $\mathbf{L}_{1-2} = \Delta \mathbf{p}$.

    (a) $\mathbf{L}_{1-2}$ represents the linear impulse of external forces only. (T or F) _____

    (b) If $\mathbf{L}_{1-2} = 0$, linear momentum is conserved. (T or F) _____

7. Consider the angular impulse-momentum principle for a rigid body:
$$(\mathbf{A}_A)_{1-2} = (\mathbf{h}_A)_2 - (\mathbf{h}_A)_1$$
    What are the two choices for point A for which this equation is valid?

    (a) _____ and (b) _____

8. When the impulse-momentum principles are applied to a system of connected rigid bodies, the impulses refer to forces that are both internal and external. (T or F) _____

B. GUIDED PROBLEMS

PROBLEM 28.1

The unbalanced 30-kg wheel rolls without slipping on the horizontal plane. When the wheel is in the position shown, determine its angular momentum about points C and A. The moment of inertia of the wheel about its mass center G is $\bar{I} = 6$ kg·m².

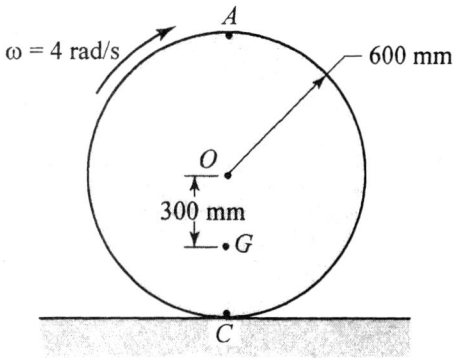

a. Comments and Analysis

- The steps that we will use in the analysis are:

    (1) Calculate $\bar{v}$, the velocity of the mass center G, and the momentum $m\bar{v}$.

    (2) Calculate the angular momentum about G.

    (3) Draw the momentum diagram for the wheel.

    (4) Using the momentum diagram, compute h_C (the angular momentum about point C) and h_A (the angular momentum about point A.)

b. Guided Solution

(1) Use the fact that C is the instant center, calculate $\bar{v}$ and the momentum:

$\bar{v} =$

$m\bar{v} =$

(2) Calculate the angular momentum about G:

$h_G = \bar{I}\omega =$

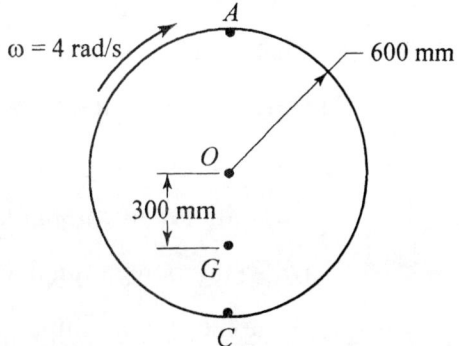

(3) Draw the momentum diagram on the figure shown.

(4) Using the momentum diagram, compute the angular momentum about C and the angular momentum about A. (Be sure to report the correct direction in your answer.)

$\circlearrowleft^+ \ h_C =$

$\circlearrowleft^+ \ h_A =$

$h_C =$ _____ ; $h_A =$ _____ **Ans.**

- -

PROBLEM 28.2

The cable attached to the 70-lb block B is wrapped around the inner radius of the drum. At the instant shown, the drum is rotating about its center of gravity G with the clockwise angular velocity $\omega_1 = 18$ rad/s. Determine the weight W of block A that will bring the system to rest in 10 seconds from the instant shown. The coefficient of kinetic friction between A and the drum is 0.2, and the central moment of inertia of the drum is $\bar{I} = 12$ slug·ft².

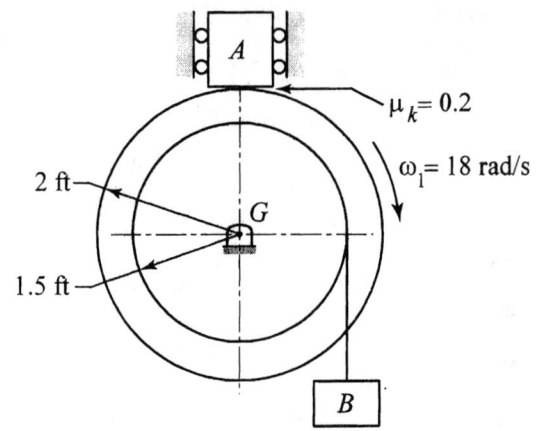

a. Comments and Analysis

- Because this problem is concerned with the change in velocity that occurs during a time interval, we will use the impulse-momentum method.
- By choosing to analyze the system of the drum and the block B, we will not be involved with the cable tension because it will be an internal force.
- The steps that we will use in the analysis are:
    (1) Draw the following three diagrams for the system of the drum and block B:
        (i) free-body diagram (FBD)
        (ii) initial momentum diagram
        (iii) final momentum diagram
    (2) Write the angular impulse-angular moment equation about G and solve for the weight W.

(continued)

b. Guided Solution

(1) On the figures below, draw the following diagrams for the system consisting of the drum and the block B:

  (i) FBD: Equilibrium analysis of the block A would reveal that the force exerted on the drum by A is simply its weight W. Be sure to show the friction force acting in its correct direction.

  (ii) Initial momentum diagram for the system (let the time be $t = 0$).

   For the drum: $\bar{I}\omega =$

   For block B: $mv =$

  (iii) Final momentum diagram for the system (let the time be $t = 10$ s). Note that this diagram will be empty because the system is at rest.

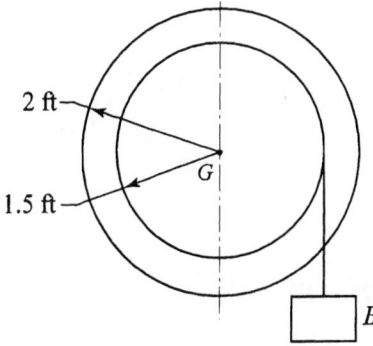 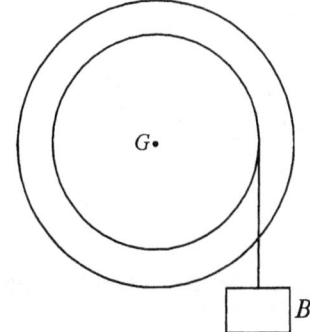 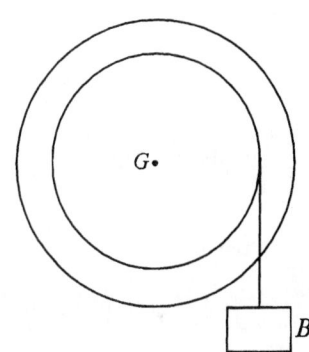

   (i) FBD of system   (ii) Initial momentum diagram   (iii) Final momentum diagram
     during motion   for system ($t = 0$)   for system ($t = 10$ s)

(2) Write the angular impulse-angular moment equation about G and solve for W.

$$(AI)_G = (h_G)_2 - (h_G)_2$$

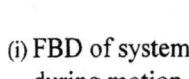

$W =$ _____ **Ans.**

Lesson 29 Rigid Body Impact

Text Reference: Articles 18.7; Sample Problems 18.12 and 18.13

Note: The analysis of rigid body impact uses the same principles that you learned when studying the impact of particle systems in Arts. 15.8 and 15.9. As a review of the concepts related to impact problems, Questions 1–4 in the following Self-Test are taken from Lesson 17. As discussed in your textbook, the coefficient of restitution is not relevant to our study of rigid body impact.

A.1 SELF-TEST

1. *(Review)*

   The contact forces caused by the collision of two particles are called
   (a) _____ forces. What notation is used to indicate these forces on free-body diagrams? (b) _____

2. *(Review)*

   What is meant by stating that the impact of two particles is *plastic*?
   _____

3. *(Review)*

   Motion for which the time of impact $\Delta t \rightarrow 0$ is called _____ motion.

4. *(Review)*

   When the time of impact is infinitesimal for two impacting blocks, the following conclusions can be made for the impact interval:

   (a) the magnitudes of impact forces are _____

   (b) the impulses of finite forces _____

   (c) the accelerations of the blocks are _____

   (d) the locations of the blocks _____

(continued)

5. Two rigid bodies collide. If the motion is assumed to be impulsive, the angular impulse equals change in angular momentum is valid for *all* points during the impact? (T or F) _____

6. The general steps in the analysis of rigid body impact are:

   (1) Draw the _____ diagram of the impacting bodies.

   (2) Draw the _____ diagram immediately before impact

   (3) Draw the _____ diagram immediately after impact

   (4) Derive and solve the appropriate _____ equations

B. GUIDED PROBLEM

PROBLEM 29.1

The 5-kg bar homogeneous bar AB is initially at rest, supported by a pin at A and a peg at B. The bar is struck by a 0.3-kg bullet C traveling at the speed $(v_C)_1 = 250$ m/s, directed as shown. The bullet becomes embedded in the bar. Determine the angular velocity ω_2 of the bar immediately after the impact. Neglect the time of impact. (Note that the subscripts 1 and 2 refer the time immediately before and immediately after impact, respectively.)

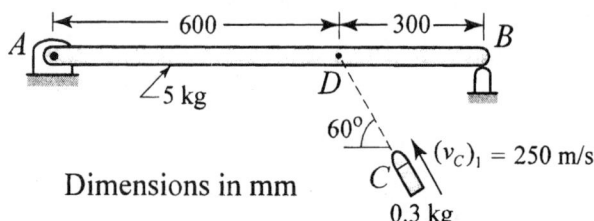

a. Comments and Analysis

- Because the time of impact is negligible, the bar and bullet are in the same position before and after impact.

- The steps that we will use in the analysis are:

    (1) Identify the impulsive forces on the FBDs of bar and bullet during impact.

    (2) Draw the following three diagrams for the system of the bar and bullet:

        (i) free-body diagram (FBD) during impact

        (ii) initial momentum diagram

        (iii) final momentum diagram

    (3) Apply the impulse-momentum principle to determine the angular velocity ω_2 of the bar immediately after the impact

b. Guided Solution

(1) The figure below shows the free-body diagrams of the bar and the bullet during impact. (The dimensions have been omitted for clarity.) Using carets over the appropriate letters, indicate which of the forces shown are impulsive forces.

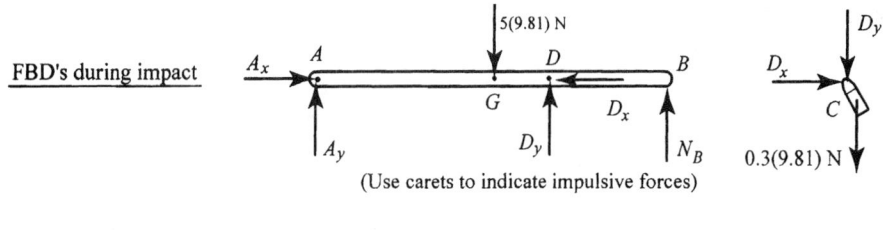

FBD's during impact

(Use carets to indicate impulsive forces)

(continued)

(2) On the figures provided, draw the following diagrams for the system:

(i) FBD: Show only the impulsive forces.

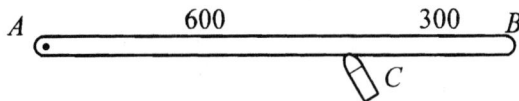

(i) FBD of system during impact showing only impulsive forces

(ii) Momentum diagram for the system before impact.

For the bullet: mv_1 =

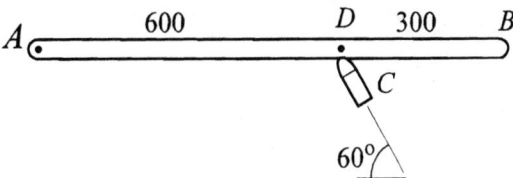

(ii) Momentum diagram for system immediately before impact

(iii) Momentum diagram for the system after impact (in terms of ω_2, CCW).

For the bullet: mv_2 =

For AB: $\bar{I} = \dfrac{mL^2}{12} =$ $\qquad\qquad m\bar{v} =$

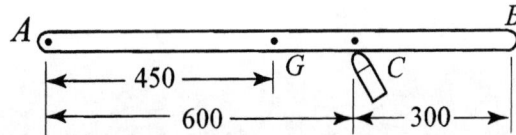

(iii) Momentum diagram for system immediately after impact

(3) Apply the impulse-momentum principle to determine ω_2.

$$(AI)_A = (h_A)_2 - (h_A)_1$$

$\omega_2 =$ _____ **Ans.**

APPENDIX A: ANSWERS TO SELF-TESTS

LESSON 1

1. particle **2.** rigid body **3.** kinematics **4.** kinetics **5.** (a) absolute; (b) relative

6. force-mass-acceleration; work-energy; impulse-momentum **7.** (a) $\left|\dfrac{d\mathbf{A}}{dt}\right|$; (b) $\dfrac{d|\mathbf{A}|}{dt}$

8. position vector **9.** (a) $\mathbf{v} = \dfrac{d\mathbf{r}}{dt}$; (b) $\mathbf{a} = \dfrac{d\mathbf{v}}{dt}$; (c) $\mathbf{a} = \dfrac{d^2\mathbf{r}}{dt^2}$ **10.** displacement

11. (a) T; (b) F **12.** speed is magnitude of velocity **13.** F/L^2

LESSON 2

1. (a) $\mathbf{r} = x\mathbf{i} + y\mathbf{j} + z\mathbf{k}$; (b) $\mathbf{v} = v_x\mathbf{i} + v_y\mathbf{j} + v_z\mathbf{k} = \dot{x}\mathbf{i} + \dot{y}\mathbf{j} + \dot{z}\mathbf{k}$;

(c) $\mathbf{a} = a_x\mathbf{i} + a_y\mathbf{j} + a_z\mathbf{k} = \ddot{x}\mathbf{i} + \ddot{y}\mathbf{j} + \ddot{z}\mathbf{k}$ **2.** particle moves in a plane

3. particle moves along a straight line **4.** eliminate all terms involving $\mathbf{i}$

5. eliminate all terms involving $\mathbf{i}$ and $\mathbf{j}$ **6.** if the particle reverses direction

LESSON 3

1. force-mass-acceleration method **2.** all forces that act on the particle

3. *ma* **4.** mass-acceleration diagram (MAD) **5.** $\Sigma F_x = ma_x$; $\Sigma F_y = 0$; $\Sigma F_z = 0$

LESSON 4

1. the motion in one direction is independent of the motion in the other two directions

2. $a_x = f_x(v_x, x, t)$; $a_y = f_y(v_y, y, t)$; $a_z = f_z(v_z, z, t)$ **3.** numerical methods

4. uncoupled

LESSON 5

1. the distance measured along the path from a fixed point

2. (a) velocity vector (b) magnitude of velocity vector (speed)

(c) acceleration vector (d) radius of curvature of the path

(e) normal component of acceleration (f) tangential component of acceleration

3. center of curvature of the path **4.** of increasing s

5. tangential component **6.** normal component **7.** normal component

8. time does not enter explicitly

LESSON 6

1. radial **2.** (a) $\mathbf{e}_R$; (b) θ **3.** (a) $\dot{R}$; (b) $R\dot{\theta}$

4. (a) $\ddot{R} - R\dot{\theta}^2$; (b) $R\ddot{\theta} + 2\dot{R}\dot{\theta}$ **5.** the radius of the circle **6.** $\mathbf{e}_z = \mathbf{k}$

LESSON 7

1. (1) free-body diagram; (2) acceleration; (3) mass-acceleration diagram;

(4) equations of motion **2.** Step 2

LESSON 8

1. $dU = \mathbf{F} \cdot d\mathbf{r}$ **2.** F_t **3.** the component of $d\mathbf{r}$ that is parallel to the force

4. (a) T; (b) F **5.** $U_{1-2} = Wh = 180(100\sin 45°) = 12\,730\,\text{lb}\cdot\text{ft}$

6. $U_{1-2} = -\frac{1}{2}k(\delta_2^2 - \delta_1^2) = -\frac{1}{2}(80)\left[(0.8)^2 - (-0.6)^2\right] = -11.20\,\text{N}\cdot\text{m}$

7. $T = \frac{1}{2}mv^2 = \frac{1}{2}(0.8)(4)^2 = 6.40\,\text{N}\cdot\text{m}$ **8.** work is a scalar quantity

9. work by the resultant force acting on a particle equals its change in kinetic energy

10. the forces that do work

LESSON 9

1. its initial and final positions   **2.** potential energy   **3.** potential energy

4. potential and kinetic   **5.** its path   **6.** T   **7.** $\frac{1}{2}k\delta^2$ **8.** Wh

LESSON 10

1. time rate of doing work   **2.** dU/dt **3.** $\mathbf{F} \cdot \mathbf{v}$ **4.** power is a scalar quantity

5. input   **6.** output   **7.** (output power / input power) x 100%

LESSON 11

1. $\int_{t_1}^{t_2} \mathbf{F}\, dt$ **2.** T   **3.** T   **4.** $m\mathbf{v}$ **5.** $d\mathbf{p}/dt$ **6.** $\Delta \mathbf{p}$ **7.** 0   **8.** T

9. (a) $\mathbf{L}_{1\text{-}2}$; (b) $U_{1\text{-}2}$; (c) $\Delta \mathbf{p}$; (d) ΔT

LESSON 12

1. $\int_{t_1}^{t_2} \mathbf{M}_A\, dt$ **2.** FLT   **3.** (a) position vector of the particle relative to point A;

(b) mass of the particle; (c) velocity of the particle   **4.** F   **5.** $(h_A)_2 - (h_A)_1$

6. a fixed point   **7.** T

LESSON 13

1. (a) absolute; (b) relative   **2.** (a) $-\mathbf{r}_{A/B}$; (b) $\mathbf{v}_A + \mathbf{v}_{B/A}$; (c) $-\mathbf{a}_{B/A}$; (d) $\mathbf{a}_A + \mathbf{a}_{B/A}$

3. F   **4.** T

LESSON 14

1. kinematic constraints   **2.** (a) kinematic; (b) position   **3.** kinematically independent

4. 3   **5.** $C - B$

LESSON 15

1. they occur in equal and opposite pairs   **2.** (a) sum of forces acting on the particle; (b) mass of the particle; (c) acceleration of the particle

3. (a) sum of external forces acting on the system; (b) total mass of the system; (c) acceleration of the mass center of the system

LESSON 16

1. (a) work done by external forces; (b) work done by internal forces; (c) change in kinetic energy of the system   **2.** spring forces and friction forces

3. T   **4.** (a) linear momentum of the sytem; (b) total mass of the system; (c) velocity of the mass center of the system   **5.** F   **6.** F

7. fixed point and mass center   **8.** (a) T; (b) F

LESSON 17

1. (a) impact; (b) a caret above a letter   **2.** velocities of the two particles are the same after impact   **3.** impulsive   **4.** (a) infinite; (b) negligible; (c) infinite; (d) do not change   **5.** (a) momentum; (b) free-body; (c) momentum

LESSON 18

1. elasticity   **2.** line of impact   **3.** (a) direct; (b) oblique

4. coefficient of restitution   **5.** 1   **6.** 0   **7.** $e = v_{sep} / v_{app}$

8. components directed along the line of impact

LESSON 19

1. all real bodies deform under load   **2.** plane   **3.** (a) translation; (b) rotation; (c) general plane motion   **4.** angular position coordinate   **5.** $\theta(t + \Delta t) - \theta(t)$

6. all lines in the body have the same angular displacement   **7.** (a) angular velocity; (b) ω **8.** (a) angular acceleration; (b) α **9.** (a) F; (b) T; (c) T

10. a circle of radius R centered on the axis of rotation

11. (a) $v = R\omega$; (b) $a_n = R\omega^2 = \dfrac{v^2}{R} = v\omega$; (c) $a_t = R\alpha$

12. (a) $\mathbf{v} = \boldsymbol{\omega} \times \mathbf{r}$; (b) $\mathbf{a}_n = \boldsymbol{\omega} \times \mathbf{v} = \boldsymbol{\omega} \times (\boldsymbol{\omega} \times \mathbf{r})$; (c) $\mathbf{a}_t = \boldsymbol{\alpha} \times \mathbf{r}$

LESSON 20

1. (a) F; (b) T; (c) T   **2.** (a) $r_{B/A}\omega$; (b) $r_{B/A}\omega^2$; (c) $r_{B/A}\alpha$; (d) $\boldsymbol{\omega} \times \mathbf{r}_{B/A}$; (e) $\boldsymbol{\omega} \times (\boldsymbol{\omega} \times \mathbf{r}_{B/A})$; (f) $\boldsymbol{\alpha} \times \mathbf{r}_{B/A}$ **3.** F   **4.** (a) 5; (b) 3

5. kinematically important point for velocity   **6.** its velocity is tangent to the path

7. (a) kinematically important; (b) relative velocity; (c) 2

8. (a) $R\omega$; (b) $R\alpha$; (c) ω; (c) α

LESSON 21

1. point on the body that has zero velocity at instant under consideration

2. when the instant center does not lie on the body   **3.** (a) F; (b) F

4. (a) perpendicular; (b) T; (c) angular velocity

LESSON 22

1. (a) 2; (b) 2; (c) 1; (d) 1; (e) 6; (f) 4; (g) ω **2.** kinematically important

3. (a) relative velocity; (b) kinematically important; (c) acceleration; (d) 2

LESSON 23

1. (a) $V_x\mathbf{i} + V_y\mathbf{j} + V_z\mathbf{k}$; (b) T **2.** (a) $V_{x'}\mathbf{i'} + V_{y'}\mathbf{j'} + V_{z'}\mathbf{k'}$; (b) $\dot{V}_{x'}\mathbf{i'} + \dot{V}_{y'}\mathbf{j'} + \dot{V}_{z'}\mathbf{k'}$

3. (a) $\boldsymbol{\omega} \times \mathbf{V}$; (b) $\boldsymbol{\omega} \times \mathbf{V}$ (c) $\dot{\boldsymbol{\omega}} \times \mathbf{V} + \boldsymbol{\omega} \times (\boldsymbol{\omega} \times \mathbf{V})$ **4.** (a) F; (b) T

(c) coincident at the time of interest; (d) $\mathbf{v}_A$; (e) (i) $\mathbf{v}_{P/\beta}$; (ii) $\mathbf{v}_{P'/A}$ **5.** (a) F; (b) T;

(c) $\mathbf{a}_A$; (d) (i) $\dot{\boldsymbol{\omega}} \times \mathbf{r}_{P/A} + \boldsymbol{\omega} \times (\boldsymbol{\omega} \times \mathbf{r}_{P/A})$; (ii) $\left(\dfrac{d^2 \mathbf{r}_{P/A}}{dt^2}\right)_{/\beta}$; (iii) $2\boldsymbol{\omega} \times \mathbf{v}_{P/\beta}$

LESSON 24

1. $\int_V r^2 dm$ **2.** $\sqrt{\dfrac{I_a}{m}}$ **3.** F **4.** axis through G that is parallel to the a-axis

5. (a) moment of inertia about the a-axis; (b) moment of inertia about the central a-axis

(c) mass of the body; (d) distance between the a-axis and the central a-axis

6. the integral of a sum equals the sum of the integrals

LESSON 25

1. (a) sum of external forces; (b) mass of the body; (c) acceleration of the mass center

2. (a) moment of inertia about axis through G; (b) angular acceleration of body;

(c) acceleration of mass center; (d) perpendicular distance from A to line of $m\bar{a}$

3. a point fixed in the body and fixed in space **4.** $\bar{I}\alpha$ **5.** 3

6. (a) inertia vector; (b) inertia couple; (c) mass center **7.** T

8. only the inertia vector acting at the mass center **9.** only the inertia couple

LESSON 26

1. T **2.** (a) $\int_{\theta_1}^{\theta_2} C\, d\theta$; (b) $C(\theta_2 - \theta_1) = C\,\Delta\theta$ **3.** (a) $C\,\Delta\theta = 10\left(\dfrac{30\pi}{180}\right) = 5.26\,\text{N}\cdot\text{m}$;

(b) $-5.26\,\text{N·m}$ **4.** $-10(20) = -200\,\text{N·m/s} = -200\,\text{W}$ **5.** (a) mass of the body;

(b) velocity of mass center; (c) moment of inertia about G; (d) angular velocity of body

6. it must be the instant center for velocities

LESSON 27

1. (a) work done by external forces; (b) work done by internal forces;

(c) change in kinetic energy of the system **2.** (a) workless;

(b) capable of doing work; (c) capable of doing work; (d) workless **3.** T

LESSON 28

1. T **2.** (a) mass of the body; (b) velocity of the mass center;

(c) moment of inertia about axis through G; (d) angular velocity of the body **3.** T

4. F **5.** center of percussion **6.** (a) T; (b) T **7.** (a) mass center; (b) fixed point

8. F

LESSON 29

1. (a) impact; (b) a caret above a letter

2. velocities of the two particles are the same after impact **3.** impulsive

4. (a) infinite; (b) negligible; (c) infinite; (d) do not change **5.** T

6. (1) free-body; (2) momentum; (3) momentum; (4) impulse-momentum

APPENDIX B: SOLUTIONS TO GUIDED PROBLEMS

PROBLEM 2.1

Given: $x = t^3 - 3t^2 - 5$ in. (x is positive to the right)

(1) $x|_{t=0} = \underline{0 - 0 - 5 = -5 \text{ in.}}$ (2) $x|_{t=4s} = \underline{(4)^3 - 3(4)^2 - 5 = 11 \text{ in.}}$

(3) [number line from -15 to 15 with points at -5 and 11, showing $\Delta \vec{r}$]

$\Delta r = \underline{16 \, \vec{\imath} \text{ in.}}$ Ans.

(4) $v(t) = \dot{x} = \underline{3t^2 - 6t \text{ in./s}}$

Time when velocity is zero: $\boxed{3t^2 - 6t = 0 \quad t = 0 \text{ and } t = 2\text{ s}}$

x when velocity is zero: $\boxed{x|_{t=0} = -5 \text{ in.} \quad x|_{t=2} = (2)^3 - 3(2)^2 - 5 = -9 \text{ in.}}$

(5) [number line showing path]

Total distance traveled = $\underline{4 + 9 + 11 = 24 \text{ in.}}$ Ans.

PROBLEM 2.2

Given $x = 103.9t$ in. and $y = -16.1t^2 + 60t$ in.

Part (a) $v_x = \dot{x} = \underline{103.9 \text{ in./s}}$ Ans. $v_y = \dot{y} = \underline{-32.2t + 60 \text{ in./s}}$ Ans.

Part (b) $v_x|_{t=0} = \underline{103.9 \text{ in./s}}$ $v_y|_{t=0} = \underline{60 \text{ in./s}}$

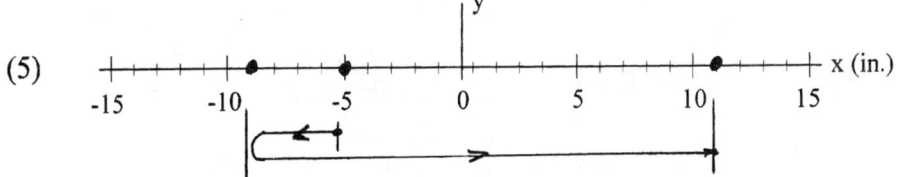

$v_0 = \underline{\sqrt{60^2 + 103.9^2} = 120 \text{ in./s}}$ Ans.

$\theta = \underline{\tan^{-1} \frac{60}{103.9} = 30°}$ Ans.

Part (c)

$$v_o = 0 = -32.2\, t_1 + 60$$
$$t_1 = \frac{60}{32.2} = 1.863 \text{ s}$$

$t_1 = \underline{1.863 \text{ s}}$

$h = y|_{t=t_1} = -16.1(1.863)^2 + 60(1.863) = \underline{55.9 \text{ ft}}$ **Ans.**

Part (d)

$$y = 0 = -16.1\, t_2^2 + 60 t_2$$
$$t_2 = 0$$
$$\text{and } t_2 = \frac{60}{16.1} = 3.727 \text{ s}$$

$t_2 = \underline{3.727 \text{ s}}$

$R = x|_{t=t_2} = 103.9(3.727) = \underline{387 \text{ ft}}$ **Ans.**

PROBLEM 3.1

Part (a) Assume block remains at rest

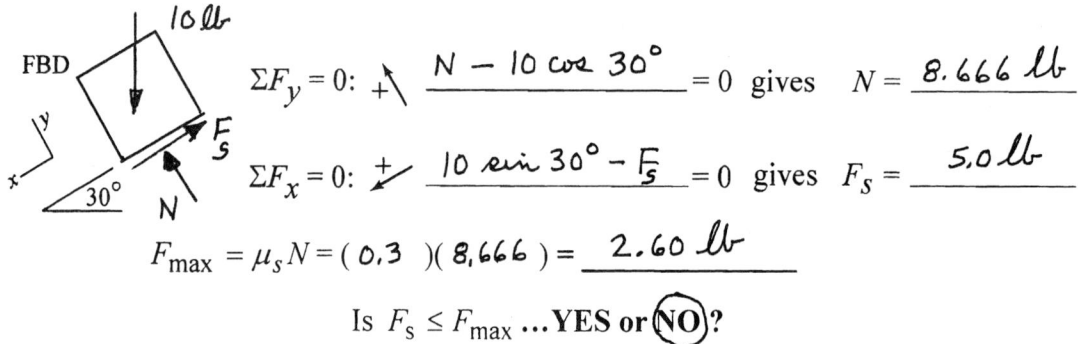

$\Sigma F_y = 0$: $\quad \underline{N - 10\cos 30°} = 0$ gives $N = \underline{8.666 \text{ lb}}$

$\Sigma F_x = 0$: $\quad \underline{10\sin 30° - F_s} = 0$ gives $F_s = \underline{5.0 \text{ lb}}$

$F_{max} = \mu_s N = (\,0.3\,)(\,8.666\,) = \underline{2.60 \text{ lb}}$

Is $F_s \le F_{max}$ ...YES or **NO**?

Block does not remain at rest

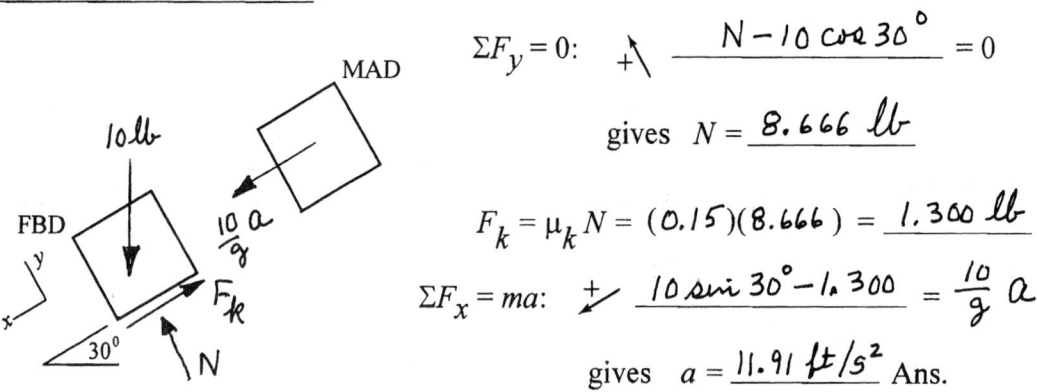

$\Sigma F_y = 0$: $\quad \underline{N - 10\cos 30°} = 0$

gives $N = \underline{8.666 \text{ lb}}$

$F_k = \mu_k N = (0.15)(8.666) = \underline{1.300 \text{ lb}}$

$\Sigma F_x = ma$: $\quad \underline{10\sin 30° - 1.300} = \frac{10}{g} a$

gives $a = \underline{11.91 \text{ ft/s}^2}$ **Ans.**

Part (b) Block sliding up the plane.

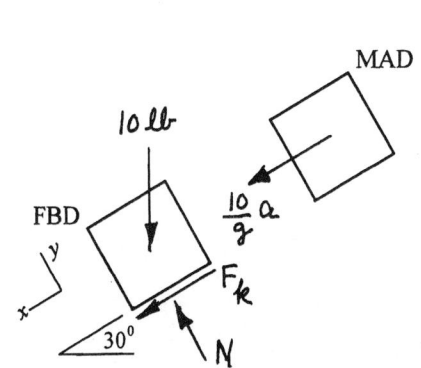

$\Sigma F_y = 0$: $\quad +\nwarrow \quad \underline{N - 10 \cos 30°} = 0$

gives $N = \underline{8.666 \text{ lb}}$

$F_k = \mu_k N = (0.15)(8.666) = \underline{1.300 \text{ lb}}$

$\Sigma F_x = ma$: $\quad +\swarrow \quad \underline{10 \sin 30° + 1.300} = \dfrac{10}{g} a$

gives $a = \underline{20.3 \text{ ft/s}^2}$ Ans.

PROBLEM 3.2

(1)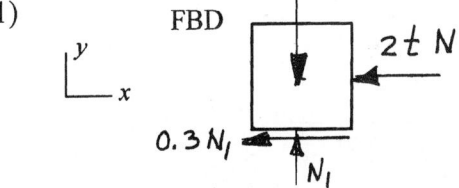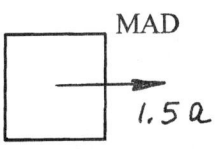

(2) $\Sigma F_y = 0$: $\quad +\uparrow \quad \underline{N_1 - 14.72 = 0} \quad$ gives $N_1 = \underline{14.72 \text{ N}}$

$\Sigma F_x = ma$: $\quad \pm\rightarrow \quad \underline{-2t - 0.3(14.72) = 1.5a}$

gives $a = \underline{-1.333 t - 2.944} \text{ m/s}^2$ Ans.

(3) $v(t) = \int a\, dt = \underline{-0.667 t^2 - 2.944 t + C_1} \text{ m/s}$

$x(t) = \int v\, dt = \underline{-0.222 t^3 - 1.472 t^2 + C_1 t + C_2} \text{ m}$

(4) (i) at $t = 0$, $v = \underline{6 \text{ m/s}}$ (ii) at $t = 0$, $x = \underline{0}$

$\boxed{\text{(ii) gives } C_2 = 0 \qquad \text{(i) gives } C_1 = 6 \text{ m/s}}$

The values are $C_1 = \underline{0} \qquad C_2 = \underline{6 \text{ m/s}}$

(5) $v(t) = \underline{-0.667 t^2 - 2.944 t + 6} \text{ m/s}$ Ans.

$x(t) = \underline{-0.222 t^3 - 1.472 t^2 + 6t} \text{ m}$ Ans.

PROBLEM 4.1

(1) FBD: box with mg downward. MAD: box with ma_y upward and ma_x to the right. Axes: y up, x right.

(2) $\Sigma F_x = ma_x$: $\xrightarrow{+}$ $\quad 0 = ma_x \quad$ gives $a_x = 0$

$\Sigma F_y = ma_y$: $+\uparrow$ $\quad -mg = ma_y \quad$ gives $a_y = -g$

(3) $a_x = 0$ $\qquad\qquad a_y = -g$

$v_x = C_1$ $\qquad\qquad v_y = -gt + C_3$

$x = C_1 t + C_2$ $\qquad\qquad y = -\dfrac{gt^2}{2} + C_3 t + C_4$

(4) 1. $t=0, x = 0$ $\qquad$ 3. $t=0, v_x = 50$ m/s
2. $t=0, y = 0$ $\qquad$ 4. $t=0, v_y = 0$

Solve for the constants:

> 1. gives $C_2 = 0$ $\qquad$ 2. gives $C_4 = 0$
> 4. gives $C_3 = 0$ $\qquad$ 3. gives $C_1 = 50$ m/s

The values are: $C_1 = 50$ m/s ; $C_2 = 0$; $C_3 = 0$; $C_4 = 0$

(5) $v_x = 50$ m/s $\qquad\qquad v_y = -gt$

$x = 50t$ m $\qquad\qquad y = -\dfrac{gt^2}{2}$

(6) Time of flight:

> t_1 is time when $y = -400$ m
> $-\dfrac{gt_1^2}{2} = -400 \qquad t_1 = 9.030$ s

$t_1 = \underline{9.03 \text{ s}}$ **Ans.**

(7) Horizontal distance:

> $d = x$ when $t = 9.030$ s
> $d = 50(9.030) = 451.5$ m

$d = \underline{452 \text{ m}}$ **Ans.**

142

PROBLEM 5.1

Part (a) $v(t) = \dfrac{ds}{dt} = 12t^2 - 10 \text{ m/s}$

$(v)_{t=3s} = 12(3)^2 - 10 = 98.0 \text{ m/s}$ **Ans.**

Part (b) $(a_n)_{t=3s} = v^2/R = (98.0)^2/15 = 640.3 \text{ m/s}^2$

$a_t = \dot{v} = 24t \text{ m/s}^2$. $(a_t)_{t=3s} = 24(3) = 72.0 \text{ m/s}^2$

$(a)_{t=3s} = \sqrt{(640.3)^2 + (72.0)^2} = 644 \text{ m/s}^2$ **Ans.**

Part (c) $(\dot{\theta})_{t=3s} = v/R = 98.0/15 = 6.53 \text{ rad/s}$ **Ans.**

$(\ddot{\theta})_{t=3s} = a_t/R = 72.0/15 = 4.80 \text{ rad/s}^2$ **Ans.**

PROBLEM 5.2

(1) (a) $\int v\, dv = \int a_t\, ds = \int (-ks + 8)\, ds$

(b) (i) $S = 0,\ v = 10 \text{ ft/s}$

(ii) $S = 6 \text{ ft},\ v = -4 \text{ ft/s}$

$\dfrac{v^2}{2} = -\dfrac{ks^2}{2} + 8S + C_1$

(c) Evaluate k and C_1.

(i) gives: $(10)^2/2 = C_1 \qquad C_1 = 50 \text{ ft/s}^2$

(ii) gives: $\dfrac{(-4)^2}{2} = -\dfrac{k(6)^2}{2} + 8(6) + 50 \qquad k = 5.00 \text{ s}^{-2}$

(d) $v^2(s) = -5S^2 + 16S + 100 \text{ ft/s}^2$

(2) $a_t = -5S + 8$ at $S=0$ $a_t = 8 \text{ ft/s}^2$ The result is $a_t = 8 \text{ ft/s}^2$

(3) $a_n = \sqrt{a^2 - a_t^2} = \sqrt{17^2 - 8^2} = 15 \text{ ft/s}^2$ The result is $a_n = 15 \text{ ft/s}^2$

(4) $\rho = v^2/a_n = 10^2/15 = 6.67 \text{ ft}$ $\rho = 6.67 \text{ ft}$ **Ans.**

PROBLEM 5.3

(a) $a_x = 0$ acceleration is vertical

(b) Answer: a_n must be directed to center of curvature

(c) $a_n = 36 \cos 60° = 18 \text{ m/s}^2$ (d) $v = \sqrt{a_n R} = \sqrt{18(2)} = 6.0 \text{ m/s}$ Ans.

PROBLEM 6.1

(a) $R = 0.4 t^2$ m $\theta = t^3/8$ rad

$\dot{R} = 0.8 t$ m/s $\dot{\theta} = 3t^2/8$ rad/s

$\ddot{R} = 0.8$ m/s^2 $\ddot{\theta} = 6t/8$ rad/s^2

(b) $R|_{t=2s} = 0.4(2)^2 = 1.60$ m $\theta|_{t=2s} = 2^3/8 = 1.0$ rad (57.3°)

$\dot{R}|_{t=2s} = 0.8(2) = 1.60$ m/s $\dot{\theta}|_{t=2s} = 3(2)^2/8 = 1.50$ rad/s

$\ddot{R}|_{t=2s} = 0.8$ m/s^2 $\ddot{\theta}|_{t=2s} = 6(2)/8 = 1.50$ rad/s^2

Part (a)

$v_R = \dot{R} = 1.60$ m/s $v_\theta = R\dot{\theta} = 1.60(1.50) = 2.40$ m/s

$v = \sqrt{(1.6)^2 + (2.40)^2} = 2.88$ m/s   The result is $v = 2.88$ m/s   Ans.

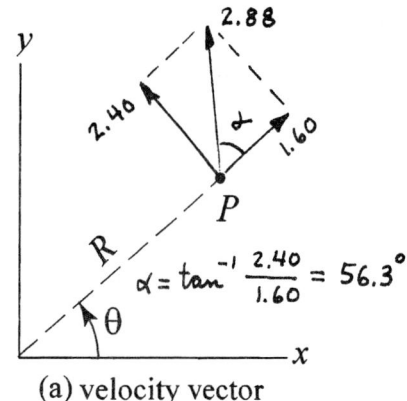
(a) velocity vector

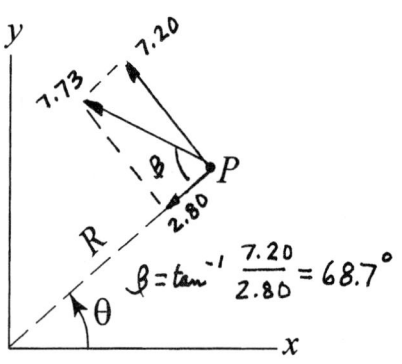
(b) acceleration vector

Part (b)

$$a_R = \ddot{R} - R\dot{\theta}^2 = 0.8 - (1.60)(1.50)^2 = -2.80 \text{ m/s}^2$$

$$a_\theta = R\ddot{\theta} + 2\dot{R}\dot{\theta} = 1.60(1.50) + 2(1.60)(1.50) = 7.20 \text{ m/s}^2$$

$$a = \sqrt{(-2.80)^2 + (7.20)^2} = 7.73 \text{ m/s}^2 \qquad \text{The result is } a = \underline{7.73 \text{ m/s}^2} \text{ Ans.}$$

PROBLEM 7.1

(1) $W = mg = 0.75(9.81) = 7.358 \text{ N}$ $a_n = v^2/R = (3)^2/0.8 = 11.25 \text{ m/s}^2$

$ma_n = \underline{0.75(11.25) = 8.438 \text{ N}}$ $ma_t = \underline{0.75\, a_t \text{ N}}$

(2)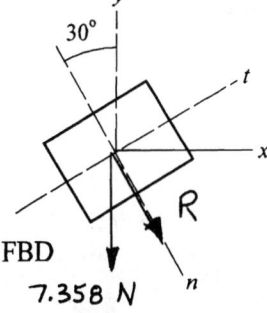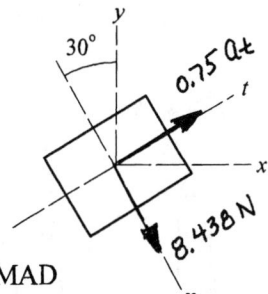

FBD   MAD
7.358 N

(3) Number of unknowns: __2__ ; number of indep. equations. of motion: __2__

(4)

$\Sigma F_n = ma_n$: ↘+ $\underline{R + 7.358 \cos 30° = 8.438}$ gives $R = \underline{2.07 \text{ N}}$ Ans.

$\Sigma F_t = ma_t$: ↗+ $\underline{-7.358 \sin 30° = 0.75\, a_t}$ gives $a_t = \underline{-4.91 \text{ m/s}^2}$ Ans.

Is the speed increasing or decreasing? __decreasing__

PROBLEM 7.2

(1) 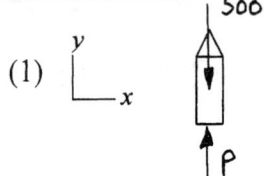 The direction of the acceleration is __vertical__

(2)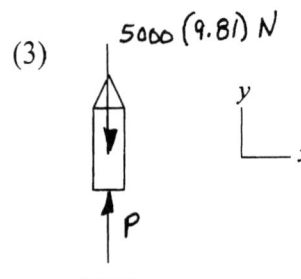

$a_\theta = R\ddot\theta + 2\dot R\dot\theta =$ __$6500(0.003) + 2(56.3)(0.02)$__

which gives $a_\theta =$ __21.75__ m/s².

$a = $ __$a_\theta/\cos 60° = 21.75/\cos 60°$__

which gives $a =$ __43.50__ m/s².

(3) FBD: $5000(9.81)$ N down, P up.   MAD: $5000(43.50)$ up.

Equation of motion:
$\Sigma F_y = ma_y$
$+\uparrow\ P - 5000(9.81) = 5000(43.50)$
$P = 267 \times 10^3$ N

$P =$ __267__ kN   **Ans.**

PROBLEM 8.1

(1) Why does N not do work? __it is always perpendicular to its path__

Weight W: $U_{A-B} = -Wy = -3(9.81)(1.5) =$ __-44.15__ N·m

40-N applied force: $U_{A-B} = Fd = 40\cos 30°(1.5) =$ __51.96__ N·m

Spring force F_s: $L_0 = 1.75$ m $L_A =$ __2__ m $L_B = \sqrt{(2)^2 + (1.5)^2} =$ __2.50__ m

$\delta_A =$ __$2 - 1.75 = 0.25$ m__ $\delta_B =$ __$2.50 - 1.75 = 0.75$ m__

$U_{A-B} = -\dfrac{1}{2}k(\delta_B^2 - \delta_A^2) = -\dfrac{1}{2}(24)(0.75^2 - 0.25^2) =$ __-6.00__ N·m

(2) $T_A =$ __0__   and $T_B =$ __$\dfrac{1}{2}mv_B^2 = \dfrac{1}{2}(3)v_B^2 = 1.5\,v_B^2$__

(3) $U_{A-B} = T_B - T_A$ $-44.15 + 51.96 - 6.00 = 1.5\,v_B^2$

$v_B = 1.10$ m/s $v_B =$ __1.10__ m/s **Ans.**

PROBLEM 8.2

(1) Why does N not do work? _it is perpendicular to its path_

Weight W: $U_{1-2} = -Wh = -16.1(20+\Delta)\sin 30° = -161 - 8.05\Delta$ lb·ft

Kinetic friction force F_k:

$N = 16.1 \cos 30° = 13.94$ lb $F_k = \mu_k N = 0.5(13.94) = 6.970$ lb

$U_{1-2} = -F_k d = -(6.970)(20+\Delta) = -139.4 - 6.970\Delta$ lb·ft

Spring force F_s: $\delta_1 = 0$ $\delta_2 = \Delta$

$$U_{1-2} = -\frac{1}{2}k(\delta_2^2 - \delta_1^2) = -\frac{1}{2}(100)(\Delta^2 - 0) = -50\Delta^2 \text{ lb·ft}$$

(2) $T_1 = \frac{1}{2}mv_1^2 = \frac{1}{2}\left(\frac{16.1}{32.2}\right)(36)^2 = 324.0$ lb·ft   and $T_2 = 0$

(3) $U_{1-2} = T_2 - T_1$

$-161 - 8.05\Delta - 139.4 - 6.970\Delta - 50\Delta^2 = 0 - 324.0$

$50\Delta^2 + 14.99\Delta - 23.6 = 0$

$\Delta = 0.553$ ft $\Delta = \underline{0.553}$ ft **Ans**

PROBLEM 9.1

(1) $(V_g)_1 = Wy_1 = \underline{0}$ and $(V_g)_2 = Wy_2 = 0.8(9.81)(0.4) = 3.139$ J

(2) $\delta_2 = \frac{W}{k} = 0.8(9.81)/300 = 0.02616$ m

$\delta_1 = \delta_2 + 0.4\text{ m} = 0.02616 + 0.4 = 0.4262$ m

$(V_e)_1 = \frac{1}{2}k\delta_1^2 = \frac{1}{2}(300)(0.4262)^2 = 27.25$ J

$(V_e)_2 = \frac{1}{2}k\delta_2^2 = \frac{1}{2}(300)(0.02616)^2 = 0.1027$ J

(3) $T_1 = 0$ and $T_2 = \frac{1}{2}mv_2^2 = \frac{1}{2}(0.8)v_2^2 = 0.4 v_2^2$

(4) $(V_g)_1 + (V_e)_1 + T_1 = (V_g)_2 + (V_e)_2 + T_2$:

$0 + 27.25 + 0 = 3.139 + 0.1027 + 0.4 v_2^2$

which gives: $v_2 = \underline{7.75 \text{ m/s}}$ **Ans.**

PROBLEM 10.1

Question 1: Answer: vectors $\vec{P}$ and $\vec{v}$ are parallel

Question 2: Answer: there is no acceleration

(a) $\dfrac{P_2}{P_1} = \dfrac{Cv_2^3}{Cv_1^3} = \dfrac{v_2^3}{v_1^3}$

$P_2 = P_1\left(\dfrac{v_2}{v_1}\right)^3 = 13.4\left(\dfrac{6}{4.5}\right)^3 = 31.77$ kW $P_2 = \underline{31.8\ kW}$ Ans.

(b) (i) Percent increase in speed = $\dfrac{6-4.5}{4.5} \times 100\%$ $\underline{33.3}$ % Ans.

(ii) Percent increase in required power = $\dfrac{31.77-13.4}{13.4} \times 100\%$ $\underline{137}$ % Ans

---

PROBLEM 11.1

(1) FBD of crate

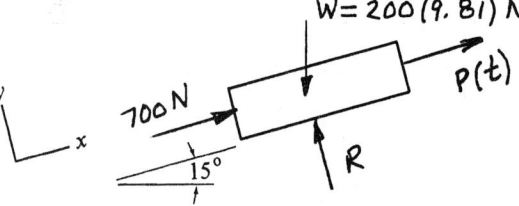

(2) *Normal force R*: $(L_{1\text{-}2})_x = \underline{\ 0\ }$ *700-N force*: $(L_{1\text{-}2})_x = \underline{700(7) = 4900\ N\cdot s}$

Weight: $(L_{1\text{-}2})_x = \underline{-200(9.81)\sin 15°(7) = -3555\ N\cdot s}$

$P(t)$: $(L_{1\text{-}2})_x = \underline{area = 150(2) + \tfrac{1}{2}(150)(2) - \tfrac{1}{2}(150)(1) - 150(2) = 75\ N\cdot s}$

(3) $(p_x)_1 = \underline{\ 0\ }$ $(p_x)_2 = \underline{200\ v_2}$

(4) $(L_{1\text{-}2}) = (p_x)_2 - (p_x)_1$

$\xrightarrow{+}\ 4900 - 3555 + 75 = 200\ v_2$

$v_2 = \underline{7.10\ m/s\ (up\ the\ plane)}$ Ans.

PROBLEM 12.1

(1) **Answer:** $(A_z)_{1-2} = 0$ because there is no moment about the z-axis

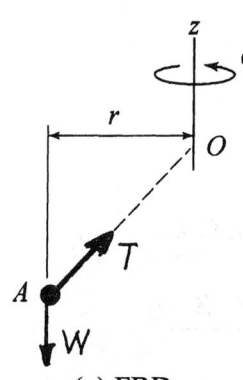
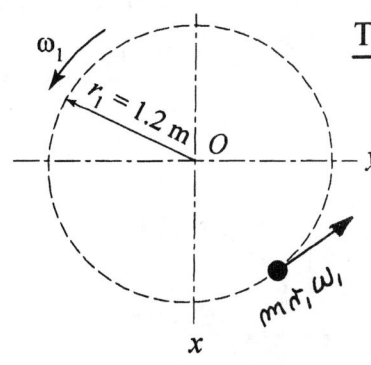
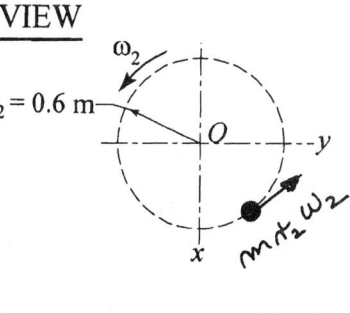

(a) FBD (arbitrary time)   (b) Momentum Diagram (time t_1)   (c) Momentum Diagram (time t_2)

(2) $(h_z)_1 = (m\, r_1\, \omega_1)\, r_1 = m\, r_1^2\, \omega_1 = m(1.2)^2(10) = 14.40\, m$ N·m·s

(3) $(h_z)_2 = m\, r_2^2\, \omega_2 = m(0.6)^2\, \omega_2 = 0.360\, m\, \omega_2$ N·m·s

(4) $(A_z)_{1-2} = (h_z)_2 - (h_z)_1$

$\circlearrowleft +$ $0 = 0.360\, m\, \omega_2 - 14.40\, m$ $\omega_2 = $ __40 rad/s__   Ans

---

PROBLEM 13.1

(1) $\mathbf{v}_A = $ __$45\, \vec{i}$ mi/h__

$\mathbf{v}_B = 35\left[(\cos 30°)\vec{i} + (\sin 30°)\vec{j}\right] = 30.31\vec{i} + 17.50\vec{j}$ mi/h

$\mathbf{v}_{B/A} = \mathbf{v}_B - \mathbf{v}_A = (30.31 - 45)\vec{i} + 17.50\vec{j}$ mi/h

The final result: $\mathbf{v}_{B/A} = $ __$-14.69\vec{i} + 17.50\vec{j}$ mi/h__   Ans.

(2) $\mathbf{r}_{B/A} = \int \mathbf{v}_{B/A}\, dt = -14.69\, t\, \vec{i} + 17.50\, t\, \vec{j} + \vec{r}_0$ mi/h

At $t = 0$, $\mathbf{r}_{B/A} = \mathbf{r}_0 = 1.5\cos 30°\, \vec{i} + 1.5\sin 30°\, \vec{j} = 1.299\vec{i} + 0.75\vec{j}$ mi

$\mathbf{r}_{B/A} = $ __$(-14.69t + 1.299)\vec{i} + (17.50t + 0.75)\vec{j}$ mi__   Ans.

PROBLEM 14.1

(1) $L = $ __$y_B + (0.6^2 + y_A^2)^{\frac{1}{2}}$__

(2) $\dfrac{dL}{dt} = 0 = $ __$v_B + \frac{1}{2}(0.6^2 + y_A^2)^{-\frac{1}{2}}(2 y_A\, v_A)$__

which gives: $v_B = \underline{\ -y_A\, \dot{y}_A / \sqrt{0.36 + y_A^2}\ }$

(3) v_B (at $y_A = 1.2$ m) = $\underline{\ -1.2(0.8)/\sqrt{0.36+(1.2)^2} = -0.716 = 0.716\ m/s\ \uparrow\ }$ **Ans.**

- -

PROBLEM 15.1

(1)
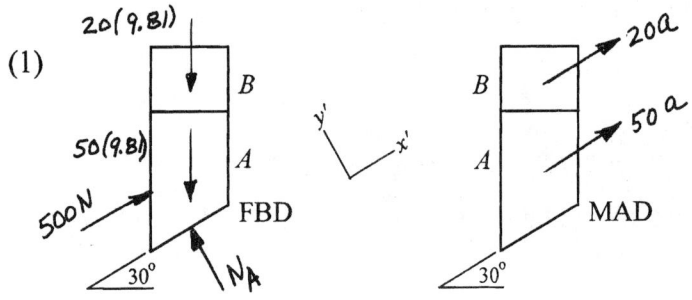

$\Sigma F_{x'} = ma_{x'}\ \nearrow\ \underline{\ 500 - 70(9.81)\sin 30° = 70a\ }$ gives $a = \underline{\ 2.238\ m/s\ }$

(2)
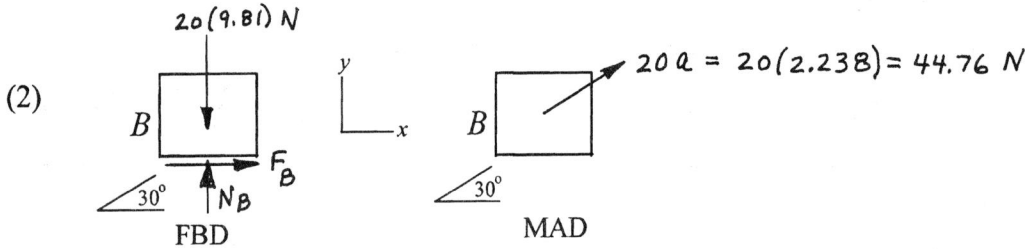

$\Sigma F_y = ma_y:\ +\uparrow\ \underline{\ -20(9.81) + N_B = 44.76\sin 30°\ }$ gives $N_B = \underline{\ 218.6\ N\ }$

$\Sigma F_x = ma_x\ \xrightarrow{+}\ \underline{\ F_B = 44.76\cos 30°\ }$ gives $F_B = \underline{\ 38.76\ N\ }$

Using $F_B = \mu N_B$ (motion impends), solve for μ:

$\mu = F_B/N_B = 38.76/218.6 = 0.177$ \hfill $\mu = \underline{\ 0.177\ }$ **ANS.**

- -

PROBLEM 16.1

(1) (a) Work done by weight of block C: $(U_{1\text{-}2})_{ext} = Wh = \underline{\ 128.8(0.8) = 103.0\ lb\cdot ft\ }$

(b) Work done by the spring.

The spring deformations are: $\delta_1 = \underline{\ 2\ ft\ }$ and $\delta_2 = \underline{\ 2.8\ ft\ }$

$(U_{1\text{-}2})_{ext} = -\dfrac{1}{2}k(\delta_2^2 - \delta_1^2) = \underline{\ -\tfrac{1}{2}(200)(2.8^2 - 2^2) = -384\ lb\cdot ft\ }$

(c) Work done by the friction force.

$\Sigma F_y = ma_y$:

$+\uparrow \quad N_A - 322 = 0 \quad$ gives $N_A = \underline{322 \text{ lb}}$

$F_A = 0.5(322) = \underline{161 \text{ lb}}$

$(U_{1\text{-}2})_{ext} = -F_A d = \underline{-161(0.8) = -128.8 \text{ lb·ft}}$

(2) $T_1 = \dfrac{1}{2} m_A (v_A)_1^2 + \dfrac{1}{2} m_C (v_C)_1^2$

$T_1 = \underline{\dfrac{1}{2}\left(\dfrac{322}{32.2}\right)(10^2) + \dfrac{1}{2}\left(\dfrac{128.8}{32.2}\right)(10^2) = 700 \text{ lb·ft}}$

(3) $T_2 = \dfrac{1}{2} m_A (v_A)_2^2 + \dfrac{1}{2} m_C (v_C)_2^2$

$T_2 = \underline{\dfrac{1}{2}(10)(v_C)_2^2 + \dfrac{1}{2}(4)(v_C)_2^2 = 7(v_C)_2^2}$

(4) $(U_{1\text{-}2})_{ext} + (U_{1\text{-}2})_{int} = T_2 - T_1$

$(103.0 - 384 - 128.8) + 0 = 7(v_C)_2^2 - 700$

$(v_C)_2 = \underline{6.44 \text{ ft/s}}$ **Ans.**

PROBLEM 16.2

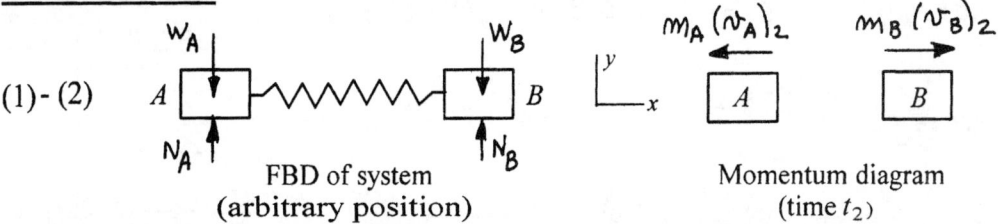

(1)-(2)   FBD of system   Momentum diagram
              (arbitrary position)   (time t_2)

(3) unknowns: $\underline{(v_A)_2}$ and $\underline{(v_B)_2}$

(4) $(L_{1\text{-}2}) = (p_x)_2 - (p_x)_1$

$\xrightarrow{+} \quad 0 = [m_B(v_B)_2 - m_A(v_A)_2] - 0$

$(v_A)_2 = \dfrac{m_B}{m_A}(v_B)_2 = \dfrac{24}{40}(v_B)_2 \qquad (v_A)_2 = \underline{0.6}(v_B)_2$ **Eq. (1)**

(5) (a) $\delta_1 = L_1 - L_0 = \underline{6-10 = -4 \text{ in}}$, $\delta_2 = L_2 - L_0 = \underline{11-10 = 1 \text{ in}}$

$(U_{1\text{-}2}) = -\dfrac{1}{2}k(\delta_2^2 - \delta_1^2) = \underline{-\dfrac{1}{2}(50)[(-4)^2 - (1.0)^2]} = \underline{375}$ lb·ft $= \underline{31.25}$ lb·in.

(b) $T_1 = \underline{0}$ \qquad (c) $T_2 = \underline{\dfrac{1}{2} m_A (v_A)_2^2 + \dfrac{1}{2} m_B (v_B)_2^2}$

(d) $(U_{1\text{-}2}) = T_2 - T_1$

$$31.25 = \tfrac{1}{2}\tfrac{40}{32.2}(v_A)_2^2 + \tfrac{1}{2}\tfrac{24}{32.2}(v_B)_2^2$$

$$0.6211(v_A)_2^2 + 0.3727(v_B)_2^2 = 31.25 \quad \text{Eq. (2)}$$

(6) $0.6211[0.6(v_B)_2^2] + 0.3727(v_B)_2^2 = 31.25$

$(v_B)_2 = 7.239\ ft/s \quad (v_A)_2 = 0.6(7.239) = 4.34\ ft/s$

$(v_A)_2 = \underline{4.34\ ft/s}$ and $(v_B)_2 = \underline{7.24\ ft/s}$ **Ans.**

PROBLEM 16.3

(1-2)

A: $mrw = 2(0.4)(90) = 72\ kg\cdot m/s$

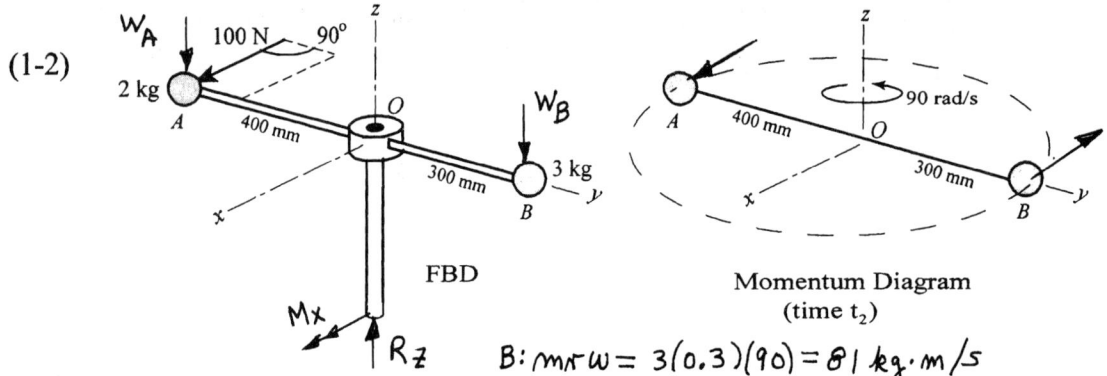

B: $mrw = 3(0.3)(90) = 81\ kg\cdot m/s$

(3) $(A_z)_{1\text{-}2} = (h_z)_2 - (h_z)_1$

$$\circlearrowleft^+ \quad 100(0.4)(t_2 - 0) = [81(0.3) + 72(0.4)] - 0$$

$t_2 = \underline{1.328\ S}$ **Ans.**

PROBLEM 17.1

(1)-(2)

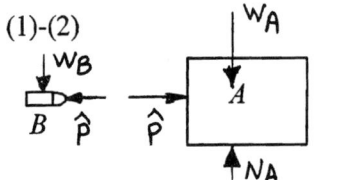
Fig. 1 FBDs of A and B during impact

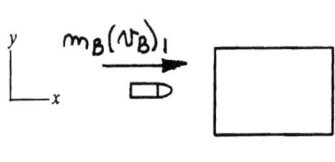
Fig. 2 Momentum diagrams for A and B before impact (Use symbols not numbers)

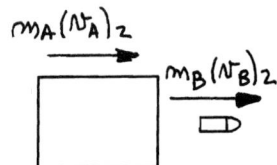
Fig. 3 Momentum diagrams for A and B after impact (Use symbols not numbers)

(3) $(L_{1\text{-}2}) = (p_x)_2 - (p_x)_1$

$$\xrightarrow{+} \quad 0 = [m_A(v_A)_2 + m_B(v_B)_2] - m_B(v_B)_1$$

$$0 = \tfrac{20}{32.2}(v_A)_2 + \tfrac{2}{16(32.2)}(500) - \tfrac{2}{16(32.2)}(2500)$$

Cancel 32.2:

$0 = 20(v_A)_2 + 62.5 - 312.5$

$(v_A)_2 = \underline{12.5\ ft/s}$ **Ans.**

PROBLEM 17.2

(1)-(2)

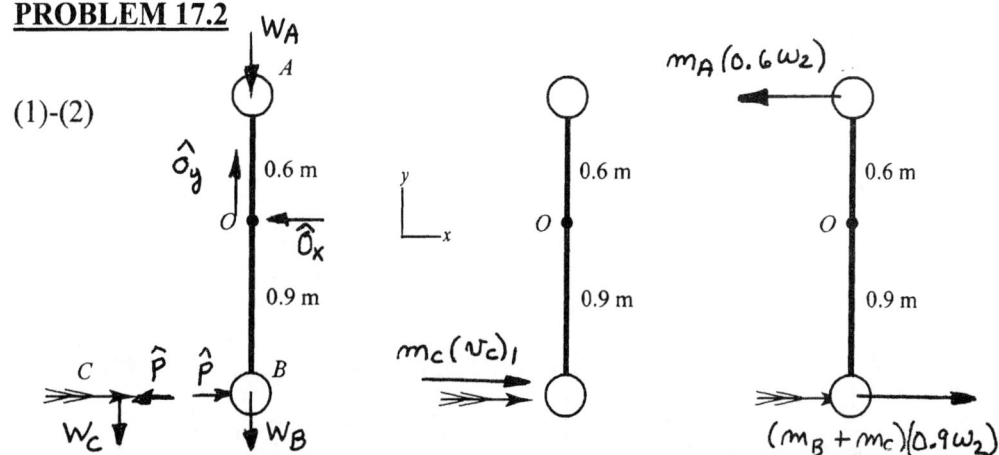

Fig. 1 FBDs during impact   Fig. 2 Momentum diagrams before impact   Fig. 3 Momentum diagrams after impact

(3) $(A_O)_{1\text{-}2} = (h_O)_2 - (h_O)_1$

$$\circlearrowright\oplus \quad 0 = \left[(m_B + m_C)(0.9\omega_2)\right]0.9 + \left[m_A(0.6\omega_2)0.6\right] - m_C(v_C)_1(0.9)$$

$$0 = \left[(1.50 + 0.075)(0.9)^2\right]\omega_2 + 0.75(0.6)^2\omega_2 - (0.075)(150)(0.9)$$

$$0 = 1.2758\,\omega_2 + 0.2700\,\omega_2 - 10.125$$

$\omega_2 = \underline{6.55\ \text{rad/s}}$ **Ans.**

PROBLEM 18.1

(1)-(2)

Fig. 1 FBD's during impact (forces acting in xy-plane only)   Fig. 2 Momentum diagram before impact   Fig. 3 Momentum diagram after impact

(3) $(L_{1\text{-}2}) = (p_x)_2 - (p_x)_1$

$$\xrightarrow{+} \quad 0 = \left[\tfrac{5}{g}(v_A)_2 + \tfrac{10}{g}(v_B)_2\right] - \left[\tfrac{5}{g}(30) - \tfrac{10}{g}(10)\right]$$

$$0 = 5(v_A)_2 + 10(v_B)_2 - 150 + 100$$

Impulse-momentum equation is: $\underline{(v_A)_2 + 2(v_B)_2 = 10}$ Eq. (1)

(4) $e = \dfrac{v_{sep}}{v_{app}}$ $0.5 = \dfrac{(v_B)_2 - (v_A)_2}{40}$

Coefficient of restitution equation is: $\underline{(v_B)_2 - (v_A)_2 = 20}$ Eq. (2)

153

(5) Eq.(1): $(v_A)_2 = 10 - 2(v_B)_2$ Eq.(2): $(v_A)_2 = (v_B)_2 - 20$

∴ $10 - 2(v_B)_2 = (v_B)_2 - 20$ gives $(v_B)_2 = 10$ ft/s

$(v_A)_2 = 10 - 2(10) = -10$ ft/s

$(v_A)_2 = \underline{10 \text{ ft/s} \leftarrow}$ and $(v_B)_2 = \underline{10 \text{ ft/s} \rightarrow}$ Ans.

PROBLEM 19.1

(1) At $t = 0$, $\theta = \underline{-2 \text{ rad}}$ 2. At $t = 0$, $\omega = \underline{3 \text{ rad/s}}$ 3. At $t = 1.0$ s, $\theta = \underline{-4 \text{ rad}}$

(2) $\omega = \int \alpha \, dt = \int (12t^2 + 2k) \, dt = 4t^3 + 2kt + C_1$ rad/s

$\theta = \int \omega \, dt = \int (4t^3 + 2kt + C_1) \, dt = t^4 + kt^2 + C_1 t + C_2$ rad

(3) 1. gives $C_2 = -2$ rad   2. gives $C_1 = 3$ rad/s

3. becomes $-4 = 1 + k + 3 - 2$ gives $k = -6$ rad/s²

$\alpha = 12t^2 + 2k = 12t^2 + 2(-6) = 12t^2 - 12$ rad/s²

$\alpha|_{t=2} = 12(2)^2 - 12 = 36$ rad/s²

$\alpha_{(t=2s)} = \underline{36 \text{ rad/s}^2 \text{ (CCW)}}$ Ans.

PROBLEM 19.2

(1) $v_{P'} = \underline{0.3(2) = 0.6 \text{ m/s} \uparrow}$ (2) $\underline{\text{they must be equal}}$

(3) $v_P = \underline{0.6 \text{ m/s} \uparrow}$ (4) $\omega_A = \underline{v_P/0.4 = 0.6/0.4 = 1.5 \text{ rad/s} \circlearrowright}$

(5) $v_C = \underline{0.2 \omega_A = 0.2(1.5) = 0.3 \text{ m/s} \downarrow}$ Ans.

(6) $\underline{\text{tangential components must be equal}}$

(7) $(a_{P'})_t = \underline{0.3(6) = 1.8 \text{ m/s}^2 \uparrow}$ (8) $(a_P)_t = \underline{1.8 \text{ m/s}^2 \uparrow}$

(9) $\alpha_A = (a_P)_t / 0.4 = \underline{\frac{1.8}{0.4} = 4.5 \text{ rad/s}^2 \circlearrowright}$ (10) $a_C = \underline{0.2(4.5) = 0.9 \text{ m/s}^2 \uparrow}$ Ans.

PROBLEM 20.1

(1) (i) _its velocity is known_   (ii) _its velocity is horizontal_

Solution I *(using scalar notation)*

(2) $\mathbf{v}_A = \mathbf{v}_O + \mathbf{v}_{A/O}$

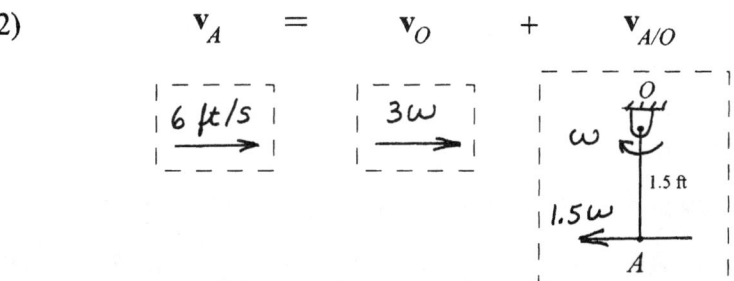

$\xrightarrow{+}\quad 6 = 3\omega - 1.5\omega = 1.5\omega$

$\omega = 4 \text{ rad/s} \circlearrowright$ 　　　　　　　$\omega = \underline{-4\vec{k}} \text{ rad/s}$ **Ans.**

Solution II *(using vector notation)*

(2) $\mathbf{v}_A = \underline{6\vec{i}} \text{ ft/s}\quad \mathbf{v}_O = \underline{3\omega\vec{i}} \text{ ft/s}\quad \omega = \underline{-\omega\vec{k}} \text{ rad/s}\quad \mathbf{r}_{A/O} = \underline{-1.5\vec{j}} \text{ ft}$

$\omega \times \mathbf{r}_{A/O} = \underline{(-\omega\vec{k}) \times (-1.5\vec{j}) = -1.5\omega\vec{i}} \text{ ft/s}$

$6\vec{i} = 3\omega\vec{i} - 1.5\omega\vec{i}$

$6 = 3\omega - 1.5\omega$

$\omega = 4 \text{ rad/s}$ 　　　　　　　　　$\omega = \underline{-4\vec{k}} \text{ rad/s}$ **Ans.**

PROBLEM 20.2

(1) (i) _they are fixed points_ (ii) _circle centered at A_

(iii) _circle centered at D_

Solution I *(using scalar notation)*

(2) $\mathbf{v}_B = \mathbf{v}_C + \mathbf{v}_{B/C}$

Eq. (a): $\xrightarrow{+}\quad 0 = -0.8\,\omega_{CD} + 0.6\,\omega_{BC}\cos 45°$

Eq. (b): $+\uparrow\quad -2.4 = 0 - 0.6\,\omega_{CD}\sin 45°$

155

Solve Eqs. (a) and (b):

(b) gives: $\omega_{BC} = 2.4/0.6 \sin 45° = 5.657$ rad/s (CCW)

(a) gives $\omega_{CD} = \dfrac{0.6 \omega_{BC} \cos 45°}{0.8} = \dfrac{0.6(5.657)\cos 45°}{0.8} = 3.00$ rad/s (CCW)

$\omega_{BC} = \underline{5.66 \text{ rad/s CCW}} \quad \omega_{CD} = \underline{3.00 \text{ rad/s CCW}}$ **Ans.**

Solution II *(using vector notation)*

(2) $\vec{\omega}_{AB} = \underline{-4\vec{k}}$ rad/s ; $\vec{\omega}_{BC} = \underline{\omega_{BC}\vec{k}}$ rad/s ; $\vec{\omega}_{CD} = \underline{\omega_{CD}\vec{k}}$ rad/s

$\vec{r}_{B/A} = \underline{0.6\vec{i}}$ m ; $\vec{r}_{C/D} = \underline{0.8\vec{j}}$ m ; $\vec{r}_{B/C} = \underline{0.6(-\sin 45°\,\vec{i} - \cos 45°\,\vec{j})}$ m

Eq. (a): $\vec{\omega}_{AB} \times \vec{r}_{B/A} = \vec{\omega}_{CD} \times \vec{r}_{C/D} + \vec{\omega}_{BC} \times \vec{r}_{B/C}$

$-4\vec{k} \times 0.6\vec{i} = (\omega_{CD}\vec{k} \times 0.8\vec{j}) + \omega_{BC}\vec{k} \times (-0.4243\vec{i} - 0.4243\vec{j})$

$-2.4\vec{j} = -0.8\omega_{CD}\vec{i} + \omega_{BC}(-0.4243\vec{j} + 0.4243\vec{i})$

Equate $\vec{j}$'s: $-2.4 = -0.4243\,\omega_{BC}$ \quad $\omega_{BC} = 5.657$ rad/s

Equate $\vec{i}$'s: $0 = -0.8\omega_{CD} + 5.657(0.4243)$ \quad $\omega_{CD} = 3.00$ rad/s

$\vec{\omega}_{BC} = \underline{5.66\,\vec{k} \text{ rad/s}} \quad \vec{\omega}_{CD} = \underline{3.00\,\vec{k} \text{ rad/s}}$ **Ans.**

PROBLEM 21.1

(a) $v_O = \underline{100(6) = 600 \text{ mm/s}}$

(b) $v_A = \underline{100\sqrt{2}\,(6) = 849 \text{ mm/s}}$

(c) $v_B = \underline{200(6) = 1200 \text{ mm/s}}$

(d) $v_D = \underline{100\sqrt{2}\,(6) = 849 \text{ mm/s}}$

(e) $v_E = \underline{\sqrt{50^2 + 100^2}\,(6) = 671 \text{ mm/s}}$
$(\theta = \tan^{-1} 0.5 = 26.6°)$

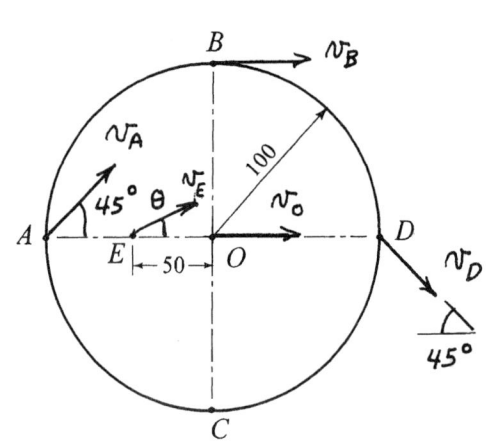

PROBLEM 21.2

(1) $v_B = r\omega_{AB} = \underline{0.6(4) = 2.4 \text{ m/s}}$ (up or <u>down</u>); (2) $v_C = r\omega_{CD} = \underline{0.8\,\omega_{CD}}$ (<u>left</u> or right)

(3)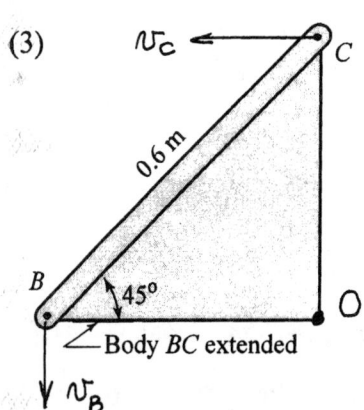

(4) $\omega_{BC} = \dfrac{v_B}{\overline{BO}} = \dfrac{2.4}{0.6\cos 45°} = 5.657 \text{ rad/s} \circlearrowleft$

$v_C = (0.6\sin 45°)\omega_{BC} = 0.6\sin 45°(5.657)$

$v_C = 2.4 \text{ m/s}$

$\omega_{CD} = \dfrac{v_C}{0.8} = \dfrac{2.4}{0.8} = 3.00 \text{ rad/s} \circlearrowleft$

$\omega_{BC} = \underline{5.66 \text{ rad/s CCW}}$; $\omega_{CD} = \underline{3.00 \text{ rad/s CCW}}$ **Ans.**

PROBLEM 22.1

Solution I *(using scalar notation)*

(2) $\mathbf{a}_B = \mathbf{a}_C + \mathbf{a}_{B/C}$

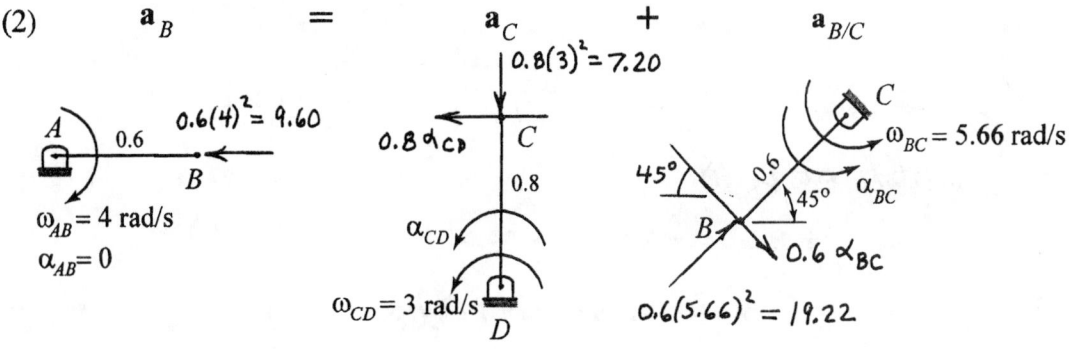

$+\uparrow\ 0 = -7.20 + 19.22\sin 45° - 0.6\,\alpha_{BC}\cos 45°$

$\alpha_{BC} = 15.06 \text{ rad/s}^2$

$\xrightarrow{+}\ -9.60 = -0.8\,\alpha_{CD} + 19.22\cos 45° + 0.6\,\alpha_{BC}\sin 45°$

Substituting $\alpha_{BC} = 15.06$ and solving gives

$\alpha_{CD} = 36.97 \text{ rad/s}^2$

$\alpha_{CD} = \underline{37.0 \text{ rad/s}^2 \text{ CCW}}$ $\alpha_{BC} = \underline{15.06 \text{ rad/s}^2 \text{ CCW}}$ **Ans.**

Solution II *(using vector notation)*

(2) $\vec\omega_{AB} = \underline{-4\vec k \text{ rad/s}}$; $\vec\omega_{BC} = \underline{5.66\vec k \text{ rad/s}}$; $\vec\omega_{CD} = \underline{3\vec k \text{ rad/s}}$

$\vec\alpha_{AB} = \underline{0}$; $\vec\alpha_{BC} = \underline{\alpha_{BC}\vec k \text{ rad/s}^2}$; $\vec\alpha_{CD} = \underline{\alpha_{CD}\vec k \text{ rad/s}^2}$

$\vec r_{B/A} = \underline{0.6\vec i \text{ m}}$; $\vec r_{C/D} = \underline{0.8\vec j \text{ m}}$; $\vec r_{B/C} = \underline{-0.4243\vec i - 0.4243\vec j \text{ m}}$

$$\omega_{AB} \times (\omega_{AB} \times r_{B/A}) = -4\vec{k} \times (-4\vec{k} \times 0.6\vec{i}) = -4\vec{k} \times (-2.4\vec{j}) = -9.60\vec{i} \text{ m/s}^2$$

$$\alpha_{AB} \times r_{B/A} = 0$$

$$\omega_{CD} \times (\omega_{CD} \times r_{C/D}) = 3\vec{k} \times (3\vec{k} \times 0.8\vec{j}) = 3\vec{k} \times (-2.4\vec{i}) = -7.20\vec{j} \text{ m/s}^2$$

$$\alpha_{CD} \times r_{C/D} = \alpha_{CD}\vec{k} \times 0.8\vec{j} = -0.8\alpha_{CD}\vec{i} \text{ m/s}^2$$

$$\omega_{BC} \times (\omega_{BC} \times r_{B/C}) = 5.66\vec{k} \times [5.66\vec{k} \times (-0.4243\vec{i} - 0.4243\vec{j})]$$
$$= 5.66\vec{k} \times (-2.402\vec{j} + 2.402\vec{i}) = 13.59\vec{i} + 13.59\vec{j} \text{ m/s}^2$$

$$\alpha_{BC} \times r_{B/C} = \alpha_{BC}\vec{k} \times (-0.4243\vec{i} - 0.4243\vec{j}) = -0.4243\alpha_{BC}\vec{j} + 0.4243\alpha_{BC}\vec{i} \text{ m/s}^2$$

$$\omega_{AB} \times (\omega_{AB} \times r_{B/A}) + \alpha_{AB} \times r_{B/A} = \omega_{CD} \times (\omega_{CD} \times r_{C/D}) + \alpha_{CD} \times r_{C/D} + \omega_{BC} \times (\omega_{BC} \times r_{B/C}) + \alpha_{BC} \times r_{B/C}$$

$$-9.60\vec{i} + 0 = -7.20\vec{j} - 0.8\alpha_{CD}\vec{i} + (13.59\vec{i} + 13.59\vec{j}) - 0.4243\alpha_{BC}\vec{j} + 0.4243\vec{i}\alpha_{BC}$$

Equate $\vec{i}$'s: $-9.60 = -0.8\alpha_{CD} + 13.59 + 0.4243\alpha_{BC}$ \quad (1)

Equate $\vec{j}$'s: $0 = -7.20 + 13.59 - 0.4243\alpha_{BC}$ \quad (2)

Equation (2) gives $\alpha_{BC} = \dfrac{13.59 - 7.20}{0.4243} = 15.06 \text{ rad/s}^2$

Substitute into Eq. (1):

$$-9.60 = -0.8\alpha_{CD} + 13.59 + 0.4243(15.06)$$

gives $\alpha_{CD} = 36.97 \text{ rad/s}^2$

$$\underline{\alpha_{BC} = 15.06\vec{k} \text{ rad/s}^2} \qquad \underline{\alpha_{CD} = 37.0\vec{k} \text{ rad/s}^2} \quad \text{Ans.}$$

PROBLEM 23.1

(1) The velocity of point A in the position shown is 15.32 ft/s, directed to the left.

Solution I *(using scalar notation)*

(2) $\mathbf{v}_P = \mathbf{v}_A + \mathbf{v}_{P'/A} + \mathbf{v}_{P/AB}$

$\xrightarrow{+}$ $(v_P)_x = -15.32 + 10\cos 40° - 5\cos 50° = -10.87$ ft/s

$+\uparrow$ $(v_P)_y = 0 - 10\sin 40° - 5\sin 50° = -10.25$ ft/s

$$v_P = \underline{-10.87\vec{i} - 10.25\vec{j}} \text{ ft/s} \quad \text{Ans.}$$

Solution II *(using vector notation)*

(2) Using vector notation, write each of the terms of the relative velocity equation

$\mathbf{v}_P = \mathbf{v}_A + \mathbf{v}_{P'/A} + \mathbf{v}_{P/AB}$.

$\mathbf{v}_A = -15.32\,\vec{i}$ ft/s

$\omega = -2\vec{k}$ rad/s ; $\mathbf{r}_{P'/A} = 5\cos 50°\,\vec{i} + 5\sin 50°\,\vec{j} = 3.214\vec{i} + 3.830\vec{j}$ ft

$\mathbf{v}_{P'/A} = \omega \times \mathbf{r}_{P'/A} = -2\vec{k} \times (3.214\vec{i} + 3.830\vec{j}) = -6.428\vec{j} + 7.660\vec{i}$ ft/s

$\mathbf{v}_{P/AB} = -5\cos 50°\,\vec{i} - 5\sin 50°\,\vec{j} = -3.214\vec{i} - 3.830\vec{j}$ ft/s

Substitute the above terms into the relative velocity equation and compute $\mathbf{v}_P$.

$\mathbf{v}_P = (-15.32\vec{i}) + (-6.428\vec{j} + 7.660\vec{i}) + (-3.214\vec{i} - 3.830\vec{j})$

$= -10.87\vec{i} - 10.26\vec{j}$ ft/s

$$\mathbf{v}_P = \underline{-10.87\vec{i} - 10.26\vec{j}} \text{ ft/s} \quad \text{Ans.}$$

PROBLEM 23.2

(1) The acceleration of point A in the position shown is 48.69 ft/s², directed to the right.

Solution I *(using scalar notation)*

(2) (units: ft/s²)

$\mathbf{a}_P = \mathbf{a}_A + \mathbf{a}_{P'/A} + \mathbf{a}_{P/AB} + \mathbf{a}_C$

- $\mathbf{a}_A$: 48.69 →
- $\mathbf{a}_{P'/A}$: $3(5) = 15$ at 40°, $5(2)^2 = 20$, 5 ft, $\alpha = 3$ rad/s², $\omega = 2$ rad/s, 50°
- $\mathbf{a}_{P/AB}$: 10 at 50°
- $\mathbf{a}_C$: $2\omega\,v_{P/AB} = 20$ at 40°, $\omega = 2$ rad/s, $v_{P/AB} = 5$ ft/s, 50°

159

$\xrightarrow{+}$ $(a_P)_x = 48.69 - 15\cos 40° - 20\cos 50° - 10\cos 50° - 20\cos 40° = 2.59 \text{ ft/s}^2$

$+\uparrow$ $(a_P)_y = 0 + 15\sin 40° - 20\sin 50° - 10\sin 50° + 20\sin 40° = -0.484 \text{ ft/s}^2$

$$\mathbf{a}_P = 2.59\vec{i} - 0.484\vec{j} \text{ ft/s}^2 \quad \text{Ans}$$

Solution II *(using vector notation)*

(2) Using vector notation, write each of the terms of the relative acceleration equation

$\mathbf{a}_P = \mathbf{a}_A + \mathbf{a}_{P'/A} + \mathbf{a}_{P/AB} + \mathbf{a}_C$.

$\mathbf{a}_A = 48.69\vec{i} \text{ ft/s}^2$

$\omega = -2\vec{k} \text{ rad/s} \quad \alpha = 3\vec{k} \text{ rad/s}^2 \quad \mathbf{r}_{P'/A} = 3.214\vec{i} + 3.830\vec{j} \text{ ft}$

$\mathbf{a}_{P'/A} = \omega \times (\omega \times \mathbf{r}_{P'/A}) + \alpha \times \mathbf{r}_{P'/A}$

$\mathbf{a}_{P'/A} = -2\vec{k} \times [-2\vec{k} \times (3.214\vec{i} + 3.830\vec{j})] + 3\vec{k} \times (3.214\vec{i} + 3.830\vec{j})$

$= -2\vec{k} \times (-6.428\vec{j} + 7.660\vec{i}) + (9.642\vec{j} - 11.490\vec{i})$

$= -12.856\vec{i} - 15.320\vec{j} + 9.642\vec{j} - 11.490\vec{i}$

$= -24.35\vec{i} - 5.678\vec{j} \text{ ft/s}^2$

$\mathbf{a}_{P/AB} = -10(\cos 50°)\vec{i} - 10(\sin 50°)\vec{j} = -6.428\vec{i} - 7.660\vec{j} \text{ ft/s}^2$

$\mathbf{a}_C = 2\omega \times \mathbf{v}_{P/AB} = 2[-2\vec{k} \times (-5\cos 50°\vec{i} - 5\sin 50°\vec{j})] = 12.86\vec{j} - 15.32\vec{i} \text{ ft/s}^2$

Substitute the above terms into the relative acceleration equation and compute $\mathbf{a}_P$.

$\mathbf{a}_P = 48.69\vec{i} + (-24.35\vec{i} - 5.678\vec{j})$
$\qquad + (-6.428\vec{i} - 7.660\vec{j}) + (12.86\vec{j} - 15.32\vec{i})$

$\mathbf{a}_P = 2.59\vec{i} - 0.478\vec{j} \text{ ft/s}^2$

$$\mathbf{a}_P = 2.59\vec{i} - 0.478\vec{j} \text{ ft/s}^2 \quad \text{Ans}$$

PROBLEM 24.1

(1) Bar AB $\bar{I}_x = \dfrac{mL^2}{12} = \dfrac{0.6}{32.2}\left(\dfrac{5^2}{12}\right) = 0.03882 \text{ slug·ft}^2 \qquad d = 2.5 \text{ ft}$

$I_x = \bar{I}_x + md^2 = 0.03882 + \dfrac{0.6}{32.2}(2.5)^2 = 0.1553 \text{ slug·ft}^2$

Bar BC $\bar{I}_x = 0$ $d = 5\,ft$

$$I_x = \bar{I}_x + md^2 = 0 + \frac{0.8}{32.2}(5)^2 = 0.6211\ slug\cdot ft^2$$

Bar CD $\bar{I}_x = \frac{mL^2}{12} = \frac{0.5}{32.2}\left(\frac{4^2}{12}\right) = 0.02070\ slug\cdot ft^2$ $d = \sqrt{5^2+2^2} = \sqrt{29}\,ft$

$$I_x = \bar{I}_x + md^2 = 0.02070 + \frac{5}{32.2}(29) = 0.4710\ slug\cdot ft^2$$

(2) $(I_x)_{assembly} = \Sigma(I_x) = 0.1553 + 0.6211 + 0.4710 = 1.247\ slug\cdot ft^2$

$(I_x)_{assembly} = \underline{1.247\ slug\cdot ft^2}$ **Ans.**

PROBLEM 25.1

(1) $a = \underline{10\cos 40° = 7.660\ ft}$
$b = \underline{5\cos 40° = 3.830\ ft}$
$c = \underline{5\sin 40° = 3.214\ ft}$

(2) Number of unknowns on FBD is $\underline{2}$. Number of unknowns on MAD is $\underline{1}$.
Total number of unknowns is $\underline{3}$. Total number of indep. eqns. of motion is $\underline{3}$.

(3) $\Sigma F_x = m\bar{a}_x$

$\xrightarrow{+}\ 8 = \frac{20}{g}\bar{a}$ $\bar{a} = 12.88\ ft/s^2$

$\Sigma(M_A)_{FBD} = \Sigma(M_A)_{MAD}$

$(\curvearrowleft +)\ 20(3.830) - N_B(7.660) + 8(2.5) = \frac{20}{g}(12.88)(3.214)$

$N_B = 9.254\ lb$

$\Sigma F_y = m\bar{a}_y$

$+\uparrow\ N_A + N_B - 20 = 0$ $N_A = 20 - N_B = 20 - 9.254 = 10.746\ lb$

$N_A = \underline{10.75\ lb\uparrow}$ $N_B = \underline{9.25\ lb\uparrow}$ $\bar{a} = \underline{12.88\ ft/s^2\rightarrow}$ **Ans.**

PROBLEM 25.2

(1)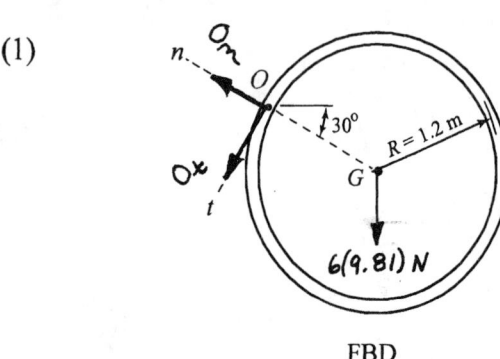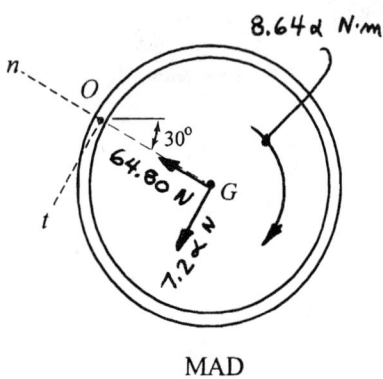

                                FBD   MAD

(2) $\bar{I} = mR^2 = 6(1.2)^2 = 8.64 \; kg \cdot m^2$

$ma_n = mr\omega^2 = 6(1.2)(3)^2 = 64.80 \; N \quad ma_t = mr\alpha = 6(1.2)\alpha = 7.2\alpha \; N$

(3) Number of unknowns on FBD is __2__. Number of unknowns on MAD is __1__.

    Total number of unknowns is __3__. Total number of indep. eqns. of motion is __3__.

(4) $\Sigma(M_O)_{FBD} = \Sigma(M_O)_{MAD}$

$\circlearrowleft^+ \quad 6(9.81)(1.2 \cos 30°) = 8.64\alpha + 7.2\alpha(1.2) \qquad\qquad \alpha = 3.540 \; rad/s^2$

$\Sigma F_n = m\bar{a}_n$

$\swarrow^+ \quad O_n - 6(9.81)\sin 30° = 64.80 \qquad\qquad\qquad\qquad O_n = 94.23 \; N$

$\Sigma F_t = m\bar{a}_t$

$\nearrow^+ \quad O_t + 6(9.81)\cos 30° = 7.2(3.540) \qquad\qquad\qquad O_t = -25.48 \; N$

$R_O = \sqrt{(94.23)^2 + (-25.48)^2} = 97.6 \; N$

                                         $\alpha = \underline{3.54 \; rad/s^2 \; \circlearrowleft}$; $R_O = \underline{97.6 \; N}$ **Ans.**

PROBLEM 25.3

(1)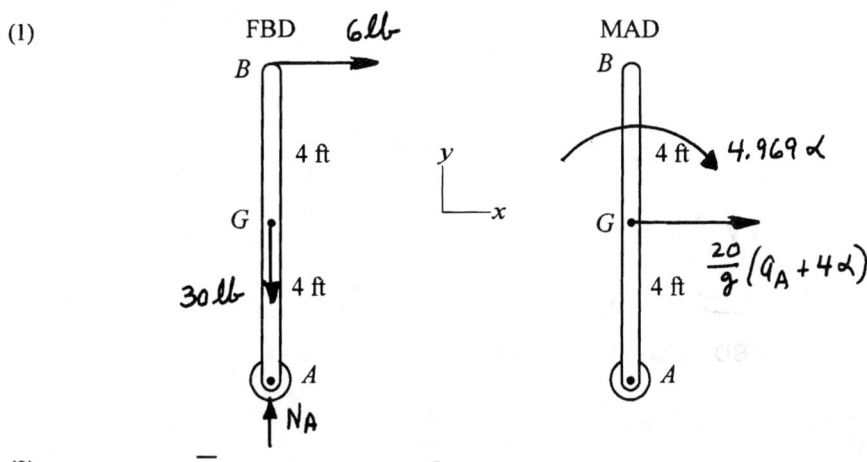

(2)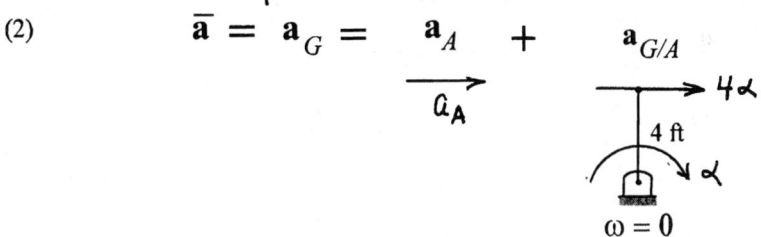

The kinematic relationship between a_G and α is $a_G = a_A + 4\alpha$

(3) $\bar{I} = \dfrac{mL^2}{12} = \dfrac{30}{32.2}\left(\dfrac{8^2}{12}\right) = 4.969$ slug·ft²

(4) Number of unknowns on FBD is __1__. Number of unknowns on MAD is __2__.
Total number of unknowns is __3__. Total number of indep. eqns. of motion is __3__.

(5) $\Sigma M_G = \bar{I}\alpha$

$\circlearrowright^+ \quad 6(4) = 4.969\alpha \qquad \alpha = 4.830 \text{ rad/s}^2 \;\circlearrowright$

$\Sigma F_x = m\bar{a}_x$

$\xrightarrow{+} \quad 6 = \dfrac{30}{32.2}(a_A + 4\alpha) = \dfrac{30}{32.2}[a_A + 4(4.830)] \qquad a_A = -12.88 \text{ ft/s}^2$

$\Sigma F_y = m\bar{a}_y$

$+\uparrow \quad N_A - 30 = 0 \qquad N_A = 30 \text{ lb}$

$\alpha = \underline{4.83 \text{ rad/s}^2 \;\circlearrowright} \quad ; a_A = \underline{12.88 \text{ ft/s}^2 \leftarrow} \quad ; N_A = \underline{30 \text{ lb} \uparrow} \qquad$ **Ans.**

- -

PROBLEM 26.1

(1) (a) Couple C Clockwise angle turned through (in radians): $\Delta\theta = 2\pi$ rad

$$(U_{1-2})_C = 1.2(2\pi) = 7.540 \text{ lb·ft}$$

(b) Weight of A Distance moved by W_A: $d_A = \frac{10}{12}(2\pi) = 5.236$ ft   (up or <u>down</u>?)

$$(U_{1-2})_{W_A} = -8(5.236) = -41.90 \text{ lb·ft}$$

(c) Weight of B Distance moved by W_B: $d_B = \frac{6}{12}(2\pi) = 3.142$ ft   (<u>up</u> or down?)

$$(U_{1-2})_{W_B} = 8(3.142) = 25.14 \text{ lb·ft}$$

(2) $(U_{1-2})_{\text{total}} = \underline{7.540 - 41.90 + 25.14 = -9.22 \text{ lb·ft}}$ **Ans.**

PROBLEM 26.2

(1)

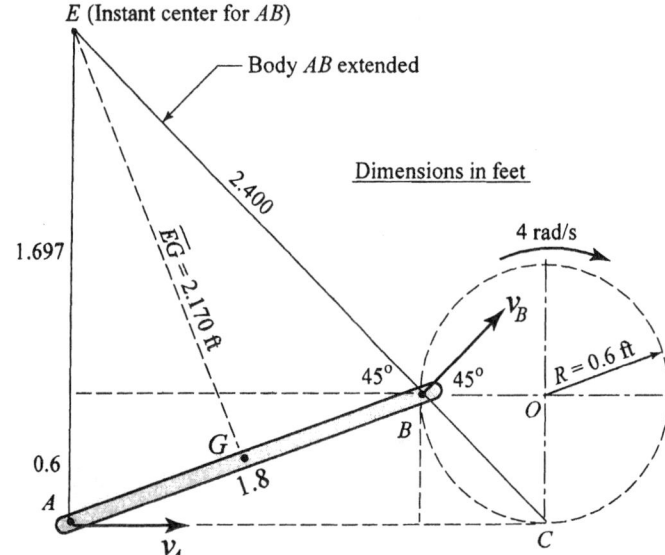

$$v_B = \sqrt{2}\, R_{cyl}\, \omega_{cyl}$$
$$= \sqrt{2}\,(0.6)(4) = 3.394 \text{ ft/s}$$

$$\omega_{AB} = v_B / \overline{BE}$$
$$= 3.394/2.400 = 1.414 \text{ rad/s}$$

$$v_A = \overline{EA}\, \omega_{AB}$$
$$= (1.697 + 0.6)(1.414)$$
$$= 3.248 \text{ ft/s}$$

(2) <u>Cylinder</u>

$$I_C = \bar{I} + md^2 = \frac{mR^2}{2} + mR^2 = \frac{3}{2}mR^2 = \frac{3}{2}\left(\frac{12}{32.2}\right)(0.6)^2 = 0.2012 \text{ slug·ft}^2$$

$$T_{cyl} = \frac{1}{2}I_C\omega_{cyl}^2 = \frac{1}{2}(0.2012)(4)^2 = 1.610 \text{ lb·ft}$$

<u>Bar AB</u>

$$I_E = \bar{I} + md^2 = \frac{mL^2}{12} + md^2 = \frac{8}{32.2}\left(\frac{1.8^2}{12}\right) + \frac{8}{32.2}(2.170)^2 = 1.237 \text{ slug·ft}^2$$

$$T_{AB} = \frac{1}{2}I_E\omega_{AB}^2 = \frac{1}{2}(1.237)(1.414)^2 = 1.237 \text{ lb·ft}$$

Block A

$$T_A = \frac{1}{2} m_A v_A^2 = \frac{1}{2}\left(\frac{2}{32.2}\right)(3.248)^2 = 0.3276 \text{ lb·ft}$$

(3) $T_{\text{total}} = T_{\text{cyl}} + T_{AB} + T_A = 1.610 + 1.237 + 0.3276 = 3.17 \text{ lb·ft}$

$T_{\text{total}} = \underline{3.17 \text{ lb·ft}}$ **Ans.**

PROBLEM 27.1

(1) Work done by the spring $L_0 = 350 \text{ mm}$

$L_1 = 500 \text{ mm}$ $\delta_1 = L_1 - L_0 = 500 - 350 = 150 \text{ mm}$

$L_2 = 500 - 500 \cos 60° = 250 \text{ mm}$ $\delta_2 = L_2 - L_0 = 250 - 350 = -100 \text{ mm}$

$$(U_{1-2})_{\text{spring}} = -\frac{1}{2}k(\delta_2^2 - \delta_1^2) = -\frac{1}{2}(800)\left[(-0.1)^2 - (0.15)^2\right] = 5.00 \text{ N·m}$$

(2) Work done by W $\Delta h = 250 - 250 \sin 60° = 33.49 \text{ mm}$

$$(U_{1-2})_W = -W\Delta h = -10(9.81)(0.03349) = -3.285 \text{ N·m}$$

(3) $T_1 = 0$ Instant center for the bar in position 2 is at point $\underline{B}$.

$$I_{\text{i.c.}} = \bar{I} + md^2 = \frac{mL^2}{12} + m\left(\frac{L}{2}\right)^2 = \frac{mL^2}{3} = \frac{10(0.5)^2}{3} = 0.8333 \text{ kg·m}^2$$

$$T_2 = \frac{1}{2}I_{\text{i.c.}}\omega_2^2 = \frac{1}{2}(0.8333)\omega_2^2 = 0.4167\,\omega_2^2$$

(4) $(U_{1-2})_{\text{spring}} + (U_{1-2})_W = T_2 - T_1$

$5.00 - 3.285 = 0.4167\,\omega_2^2 - 0$

$\omega_2 = 2.03 \text{ rad/s}$ $\omega_2 = \underline{2.03 \text{ rad/s}}$ **Ans.**

PROBLEM 28.1

(1) $\bar{v} = 0.3(4) = 1.2 \text{ m/s}$

$m\bar{v} = 30(1.2) = 36 \text{ N·s}$

(2) $h_G = \bar{I}\omega = 6(4) = 24 \text{ N·m·s}$

(4) ↻ $h_C = 24 + 36(0.3) = 34.8 \text{ N·m·s}$

↻ $h_A = 24 - 36(0.9) = -8.4 \text{ N·m·s}$

$h_C = \underline{34.8 \text{ N·m·s}} \circlearrowright$ $h_A = \underline{8.4 \text{ N·m·s}} \circlearrowleft$ **Ans.**

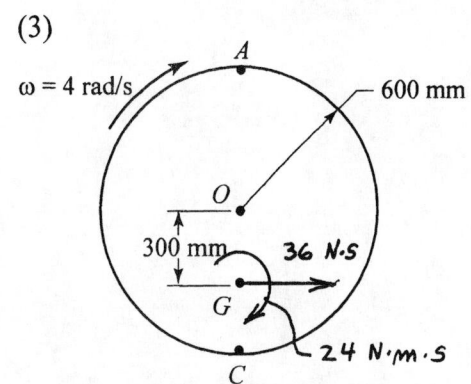

PROBLEM 28.2

(1) (ii) For the drum: $\bar{I}\omega = 12(18) = 216$ lb·ft·s

For block B: $mv = \dfrac{70}{32.2}(1.5)(18) = 58.70$ lb·s

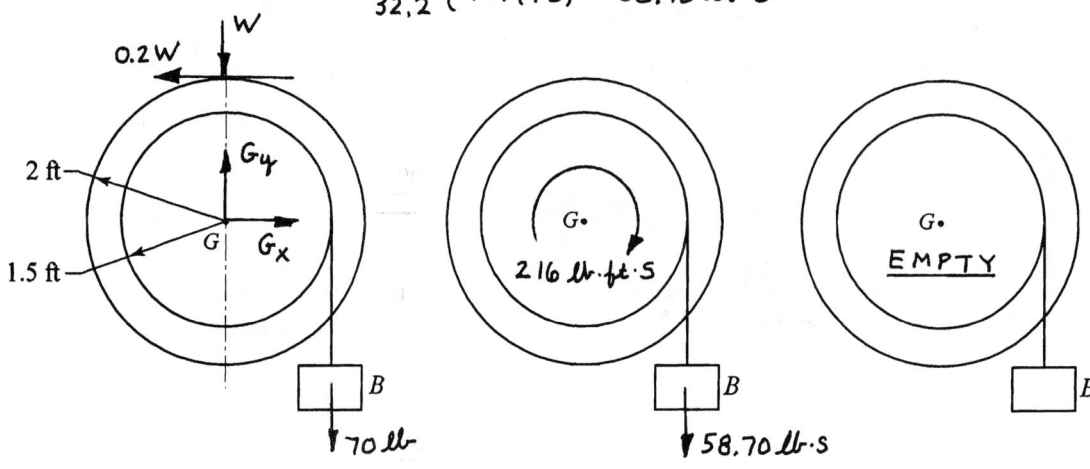

(i) FBD of system during motion

(ii) Initial momentum diagram for system ($t = 0$)

(iii) Final momentum diagram for system ($t = 10$ s)

(2) $(AI)_G = (h_G)_2 - (h_G)_1$

$\curvearrowright^+ \quad [70(1.5) - 0.2W(2)](10) = 0 - [216 + 58.70(1.5)]$

$1050 - 4W = -304.05$

$W = 338.5$ lb $W = \underline{\quad 339 \text{ lb} \quad}$ Ans.

PROBLEM 29.1

(1)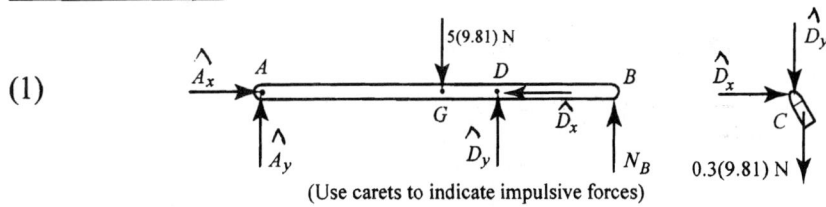

(Use carets to indicate impulsive forces)

(2)

(i)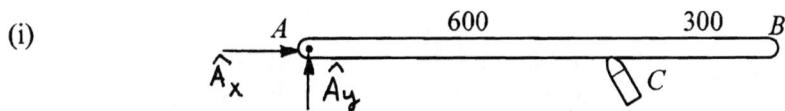

(i) FBD of system during impact showing only impulsive forces

(ii) For the bullet: $mv_1 = 0.3(250) = 75\ N\cdot s$

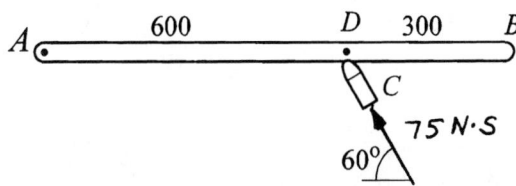

(ii) Momentum diagram for system immediately before impact

(iii) For the bullet: $mv_2 = 0.3(0.6\ \omega_2) = 0.18\ \omega_2$

For AB: $\bar{I} = \dfrac{mL^2}{12} = \dfrac{5(0.9)^2}{12} = 0.3375\ kg\cdot m^2 \quad m\bar{v} = 5(0.45\omega_2) = 2.25\ \omega_2$

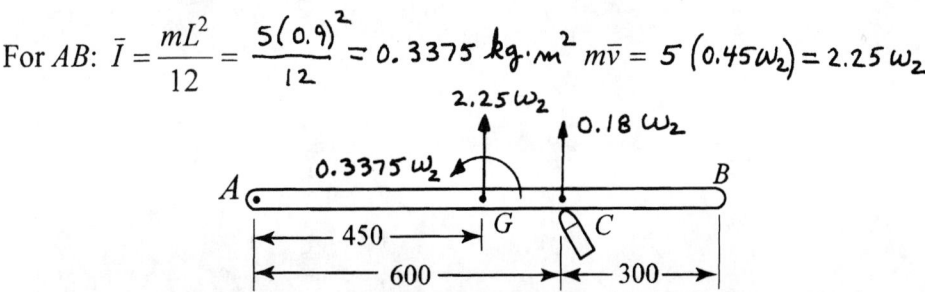

(iii) Momentum diagram for system immediately after impact

(3) $(AI)_A = (h_A)_2 - (h_A)_1$

↶⊕ $0 = [0.3375\ \omega_2 + 2.25\ \omega_2(0.45) + 0.18\ \omega_2(0.6)] - 75\sin 60°(0.6)$

Solving gives $\omega_2 = 26.7\ rad/s$

$\omega_2 = \underline{26.7\ rad/s}$ **Ans.**